Design of
Blow Moulds

Design of
Blow Moulds

R.C. Batra
German Advisor (Retd.)

CBS

CBS PUBLISHERS & DISTRIBUTORS PVT. LTD.
New Delhi • Bangalore • Pune • Cochin • Chennai (India)

ISBN : 978-81-239-1495-4 (PB)
       978-81-239-1499-2 (HB)

First Edition : 2007
Reprint : 2010

Published by Satish Kumar Jain and produced by V.K. Jain for
CBS Publishers & Distributors Pvt. Ltd.,
CBS Plaza, 4819/XI Prahlad Street, 24 Ansari Road, Daryaganj,
New Delhi - 110002, India.   • Website: www.cbspd.com
e-mail: delhi@cbspd.com, cbspubs@vsnl.com, cbspubs@airtelmail.in
Ph.: 23289259, 23266861, 23266867   • Fax: 011-23243014

Branches:
• *Bangalore:* Seema House, 2975, 17th Cross, K.R. Road,
  Bansankari 2nd Stage, Bangalore - 560070   Ph.: 26771678/79
  Fax: 080-26771680   • e-mail: bangalore@cbspd.com
• *Pune:* Bhuruk Prestige, Sr. No. 52/12/2+1+3/2,
  Narhe, Haveli (Near Katraj-Dehu Road by Pass), Pune-411051
  Ph.: +91-20-32404169   • Fax: 020-24464059
  e-mail: pune@cbspd.com
• *Cochin:* 36/14, Kalluvilakam, Lissie Hospital Road,
  Cochin - 682018, Kerala • e-mail: cochin@cbspd.com
  Ph.: 0484-4059061-65   • Fax: 0484-4059065
• *Chennai:* 20, West Park Road, Shenoy Nagar, Chennai - 600030
  e-mail: chennai@cbspd.com   Ph.: 044-26260666-26202620
  Fax: 044-45530020

Printed at :
India Binding House, Noida, UP

Preface

Extrusion blow moulding is one of the major processes for mass production of containers, hollow bodies and industrial articles out of thermoplastics. Inexplicably, the design principles for blow moulds, which wield considerable influence on quality and productivity, have received scant attention. This book is an attempt to combine and compile knowledge and experience of raw material producers, machine manufacturers, blow moulders, designers and mould makers on this subject.

It has been endeavoured to provide the mould designers with practical data, well- tested methods and down to earth tips rather than complicated formulae, requiring hard to get data. The sum total of "theoretical considerations and calculations" has been presented in practical form of tables and charts to facilitate direct application in day-to-day work of mould design. The text has been illustrated with appropriate diagrams for easy comprehension

I am greatly indebted to numerous German firms and institutions viz. BASF, Hoechst, Kautex, Bekum and IKV-Aachen among others who allowed me to draw freely on their publications and reproduce diagrams and pictures.

I would also like to acknowledge the encouragement, guidance and assistance received from friends: Mr Bellinghausen, Mr. Goswami and Mr. Satya Prakash, to name a few. As a mark of appreciation and gratitude, I dedicate this monograph to them.

Introduction

Tools play an important role in all manufacturing processes. Their influence is, however, more far-reaching in the processes involving plastics. They not only mould the product to its final shape but also influence the productivity, quality, life and the cost. A well-designed tool produces the goods economically, efficiently and consistently with minimum maintenance.

To create a tool with above attributes, a mould designer is required to have a thorough knowledge, not only of the design principles but also of the material to be processed, the method of processing and the machinery therefor, the materials and methods for manufacturing the tool, the cost aspects, the pitfalls and their remedies. These considerations have provided the guidelines for selection and extent of the contents.

The blow moulding field has been going through an exponential development, ever since its importance for the technical sector, especially the automobile industry, has been recognised. A good designer must keep track of the latest developments in order to turn out state of art tools.

It is hoped that this monograph will be found useful by designers, mould makers, processors and students of the plastics technology.

Contents

Plastics

Plastics is a generic term applying to a wide range of synthetic materials, produced either by modification of some natural matter like cellulose or by chemical synthesis of elements such as carbon, hydrogen, chlorine, nitrogen, oxygen, silicon, sulphur etc. in long chains. Some of the plastics found in the nature are shellac, rubber and wax. Before we go into the details of individual plastics, it will be advisable to go through some general aspects, which apply to all of them.

Plastics are man-made materials containing, as their essential ingredient, long molecular chains of elements like carbon, hydrogen, oxygen, nitrogen etc., made of recurring building blocks or monomers (Fig. 1.1). While solid in the finished state, at some stage during their manufacturing, they are soft enough to be formed into various shapes, most commonly through the application, either singly or together, of heat and pressure. Plastics expand when heated and contract upon cooling.

Thermosetting Plastics or Thermosets harden or "cure" under heat into a permanent shape because of an irreversible chemical reaction resulting in permanent cross-linking of molecular chains (Fig. 1.2). They do not soften or remelt if heated again but disintegrate or char.

Fig. 1.1.

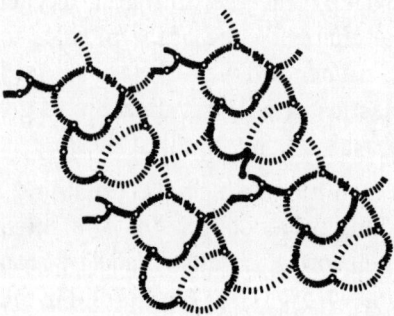

Fig. 1.2.

Thermoplastics undergo a physical change upon heating. At some stage, they soften under heat and harden again when cooled. In the softened or molten state, they can be made to flow under pressure to assume the shape of the receptacle or **mould** into which they are forced. The shape is retained upon cooling but the plastics can be reheated and given another shape a number of times. In the molten state, the molecular chains can slide past each other more easily and the plastics 'flows'. There is no cross-linking of chains; the change in state is physical and reversible. Their chemical structure remains unaltered in cold as well as hot state.

Basically, only thermoplastics are used for blow moulding

There are two main varieties of thermoplastics:

The Amorphous Thermoplastics. The molecular chains of these plastics are not arranged in an orderly fashion (Fig.1.3). When melted, their volume increases upto 15%.

- They are hard and rigid.
- Their linear shrinkage is relatively small and varies between 0.3-0.8%.
- Amorphous thermoplastics are transparent in their basic form.
- Their specific gravity ranges from 1.05 to 1.4
- Amorphous thermoplastics do not have a distinct melting point but start softening at some stage, called glass temperature, when heated.
- The most common amorphous thermoplastics are PS, SAN, ABS, PMMA, and PC etc.

The Semi-Crystalline Thermoplastics have part of their chains arranged in a parallel mode. The rest is disorderly like those of the amorphous plastics (Fig. 1.4).

- They are more flexible than the amorphous thermoplastics.
- Their volume increases by about 30% or more upon

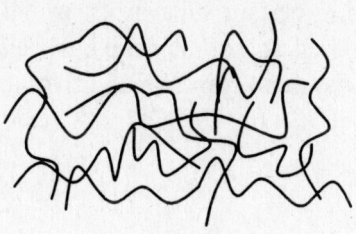

Fig. 1.3.

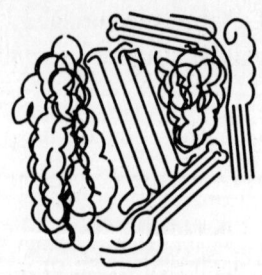

Fig. 1.4.

melting as the closely packed macromolecules get disorderly
- Their linear shrinkage is relatively large; it ranges from 1.5 to 5%
- In their natural state, they are translucent to opaque.
- Their specific gravity lies between 0.9 and 1.3
- While melting, they require latent heat. They have a distinct melting point.
- The most common semi-crystalline thermoplastics are: LDPE, HDPE, PP, PA, POM, PET etc.

Liquid Crystal Polymers, though crystalline, form a special class of thermoplastics. They possess rod like crystals, even in the molten state, and have properties, which make them an ideal replacement of metals in many cases.

Additives. In order to improve the processability of polymers and to modify some characteristics or to incorporate some missing properties, certain substances, usually chemicals, are added to them. They do not alter the basic chemical character of the plastics. The most common additives are:

- Plasticisers
- Stabilisers
- Lubricants
- Impact modifiers

- Fire retardants
- UV-stabilisers
- Anti-static agents
- Anti-slip agents
- Blowing agents
- Pigments
- Fillers and reinforcements
- Catalysers
- Accelerators
- Special purpose additives

Reinforced Plastics. To enhance certain properties of plastics, reinforcing components are physically mixed into them. Glass fibres are the most common addition. The other reinforcing ingredients are glass beads, mineral fibres, talcum etc. For special applications, steel fibres have also been incorporated. The reinforcing components show beneficial effect in case of many physical properties. There are a few adverse effects also.

There is an **increase** in the values of:

Modulus of elasticity, creep resistance, tensile strength, heat conductivity, high temperature resistance, specific gravity, impact strength below the freezing point.

Following attributes experience a **decrease**:

Coefficient of thermal expansion, cooling time as the mineral have lower specific heat and better heat conductivity, mould shrinkage, impact strength at room temperature, flowability and weld line strength.

Polyblends are a mixture of two or more polymers having different properties. The mixtures are created with the following aims:

- Improving polymer processability
- Enhancing the physical and mechanical properties
- Developing tailor made materials

- Reducing the material cost by mixing a low cost material with an expensive one
- Recycling the mixed scrap

As most of the blends are immiscible, phase separation occurs during processing. In order to overcome this difficulty, special coupling agents including some copolymers called compatibilizers are added to the mixture.

Elastomers are polymers with reversible elastic properties. They can be stretched to a manifold length at room temperature. On removal of the stress, they resume their shape and length (Fig. 1.5).

Rubber is a natural elastomer.

Synthetic elastomers can be thermoplastic or thermosetting.

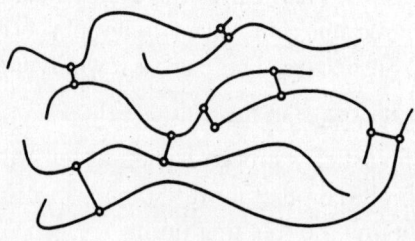

Fig. 1.5.

The effect of molecular structure on properties

The physical and mechanical properties of plastics depend upon the shape of their macromolecules and the forces, which hold them together.

The Thermosetting Plastics have three-dimensional macromolecules. The binding force among the molecules is the chemical valency, which is very strong. This explains why the processed thermosets are hard and do not flow under heat and pressure.

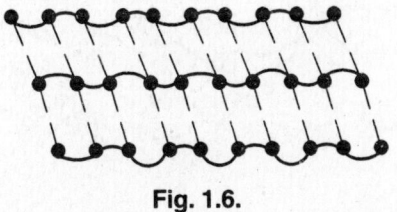

Fig. 1.6.

The Thermoplastics have linear macromolecules, consisting of monomers bound by the strong force of their valencies (Fig. 1.6). The molecules of some thermoplastics may have side branches too. The holding force between the macromolecules is the co-ordinate bond or an electrostatic force, which is weak and gets weaker with increase in distance, temperature and pressure. The linear structure permits the polymer segments to get closer to each other than the branched structure so that the former is denser (Fig. 1.7). The branched side chains, on the other hand, will keep the main polymer backbones further apart from each other (Fig. 1.8). The longer the chain, the greater is the effect. The denser polymer would have higher **Specific Gravity**.

Out of the same reason, the polymer with branched chains and more space between chains would have higher **Permeability** to gases.

Linear materials with closely packed chains will have higher intermolecular forces and therefore higher **Tensile Strength**.

Likewise, the linear materials being closely packed, will be

Fig. 1.7.

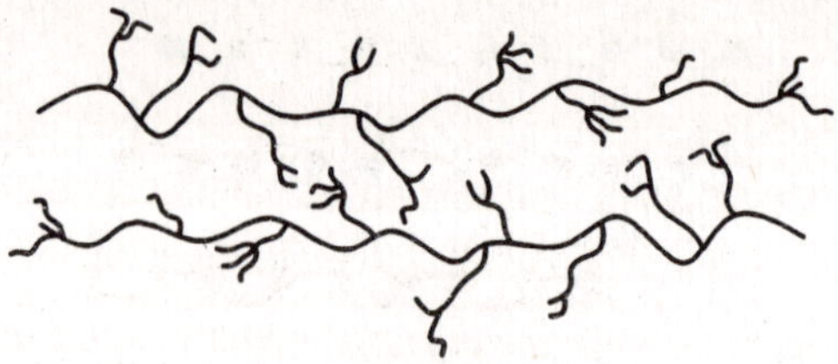

Fig. 1.8.

stiffer because of less room for bending of the backbones. Their **Modulus of Elasticity** will be higher.

As the molecular forces binding a high density material are higher, it takes a higher amount of heat energy to separate the linear configuration. The **Heat Distortion Temperature** will be higher.

Hardness, which can also be defined as resistance to penetration, will be higher for materials having closely packed chains.

Creep Resistance or the resistance to distorting forces too is higher with materials of linear chains because of higher intermolecular forces.

Flowability is lower for linear materials in case of stronger molecular attraction. The flow is opposed by the cohesive forces.

Branched chains have space between themselves. Hence these materials are more easily **compressible** than the ones with straight chains.

The **Impact Strength** is, however, better with branched materials, which are more flexible.

Semi-crystalline Thermoplastics have a part of their molecules lying closer together in an orderly fashion. Their physical and mechanical properties are closely linked not only with their crystalline structure but also with the extent and the form of crystallinity. The compactness as a result of closeness results in better cohesion between the molecular chains, which increases the density, the stiffness, the heat resistance etc. The effect is similar to that of the unbranched chains. It may be summed up that:

- The more the crystallinity, the higher the density.
- The more the crystallinity, the greater the stiffness as the closely packed crystals have less space to rotate.
- The more the crystallinity, the higher the tensile strength as more force is needed to break the intimate bonds.
- The more the crystallinity, the more the hardness (resistance to penetration) because of the closely packed molecular chains.
- The more the crystallinity, the less the permeability to gases as there are less gaps between the chains.
- The more the crystallinity, the higher the softening temperature as more heat energy is needed to overcome the cohesive forces.
- The more the crystallinity, the more the volume expansion on melting as the freed chains get disorderly.
- The more the crystallinity, the greater the shrinkage on cooling. Ordered molecules occupy less space.
- The more the crystallinity, the less the impact strength. Crystals propagate cracks.
- The more the crystallinity, the less the resistance to stress cracking.
- Difference in crystallinity in different sections results in more warpage.

The extent of crystallinity in a blow-moulded article depends upon mould temperature and the cooling time.

Flowability of thermoplastics is directly related to the length of the molecular chains. Shorter chains can slide past one another more easily than the longer ones. This is why, the material grades with longer chains or higher molecular weights are more viscous. Their melts have more strength.

Molecular Weight is the weight of all building blocks or monomers in the chain of a thermoplastics. All chains in a plastics are, however, not of the same length. Therefore a particular material grade is known by its average molecular weight. The grades with higher average molecular weight are distinguished by better creep resistance, chemical resistance and impact strength due to an increase in entanglement and intermolecular attraction between the adjacent chains. However, the flowability suffers because of the same reasons.

Melt Flow Index (MFI) is a measure of relative viscosity or flowability of thermoplastics. It is the weight of the melt in grams, extruded through a standard die at specified temperature under a specified force, in 10 minutes (Fig. 1.9). The MFI is expressed as a number. The higher the number, better is the flowability. It holds good only for individual thermoplastics. Melt flow indices of different thermoplastics cannot be compared with one another.

Flowability of PVC cannot be measured by the above process because of the thermal instability of the melt. Measure of

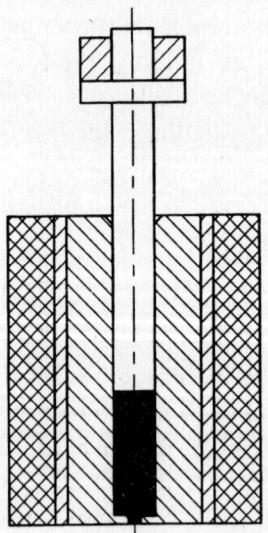

Fig. 1.9.

the flowability of PVC is the so-called K value. A higher K-value indicates higher viscosity or lower flowability.

Thermoplastics for the extrusion process should have a high viscosity. Their MFI should be low. The extrusion grade PVC has a high K-value.

Formation of plastics from crude oil to polymers

The crude oil contains hydrocarbons in various combinations like gases, light and heavy oils, petrol, coal tar etc. These are separated by fractional distillation. Heavy benzene or naphtha is the raw material for plastics. By application of heat (850°C) in presence of catalysts, its molecule is cracked into smaller unsaturated compounds like Ethylene and Ethane (Fig. 1.10).

Ethylene, a gas, is separated by liquefying the mixture. It forms basis for many plastics as well as other compounds. C_2H_6 is its monomer or one building block. By **polymerisation**, which is formation of a long chain or a so-called macromolecule consisting of many individual building blocks or monomers, it is converted into Polyethylene, which is a solid plastics. **Polymerisation** can be represented graphically by the following diagram (Fig. 1.11):

As mentioned before, the monomer of Ethylene (Fig.1.12) forms the basis for many other plastics. If one of the four Hydrogen atoms is replaced by a CH_3 group, it yields a monomer of Propylene. Through the process of polymerisation, we get the plastics known as Polypropylene (Fig. 1.13).

Likewise, substitution of one atom of Hydrogen in the monomer by one atom of Chlorine converts it into the monomer of Vinyl Chloride (Fig. 1.14).

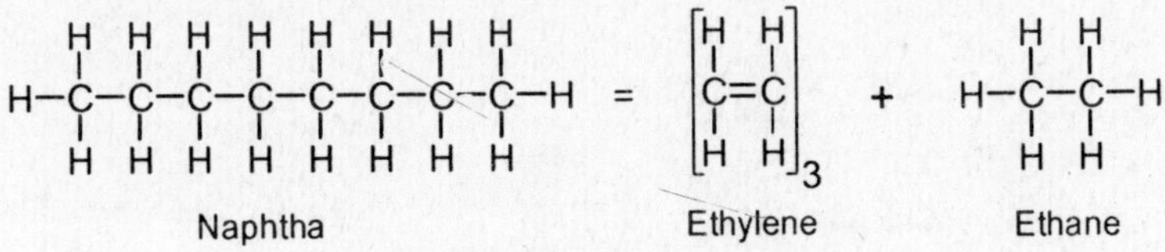

Naphtha   Ethylene   Ethane

Fig. 1.10.

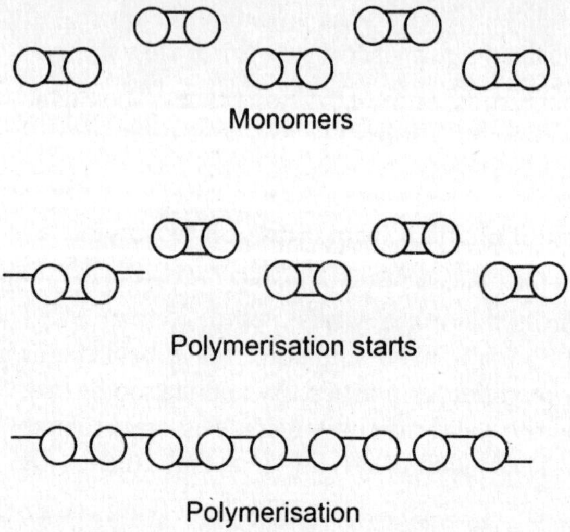

Monomers

Polymerisation starts

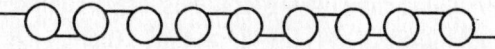

Polymerisation

Fig. 1.11.

The monomer of Acrylonitrile is formed by replacing one atom of hydrogen by a CN group (Fig. 1.15).

Similarly, a Benzol ring in place of one Hydrogen atom yields a molecule of Styrene (Fig. 1.16).

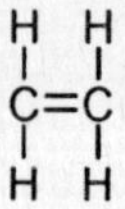

Ethylene

Fig. 1.12.

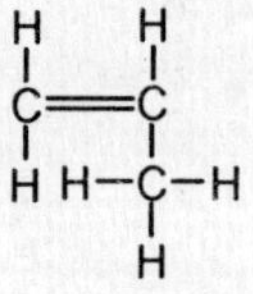

Propylene

Fig. 1.13.

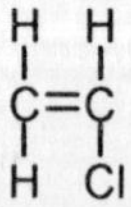

Vinyl Chloride

Fig. 1.14.

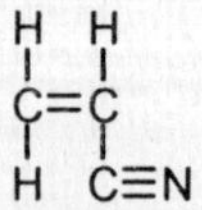

Acrylonitrile

Fig. 1.15.

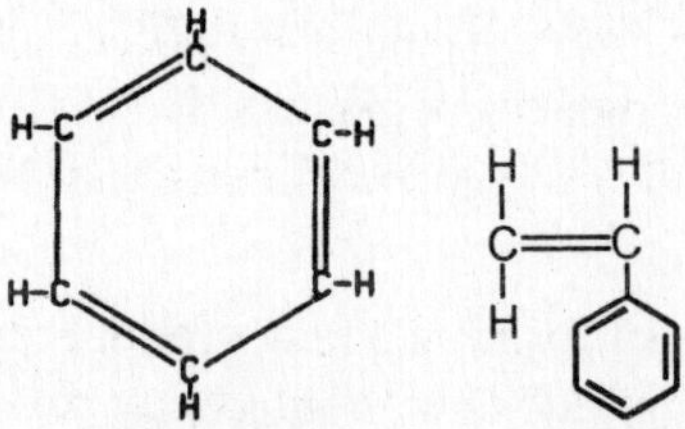

Benzol Ring   Styrene   Tetrafluoroethylene

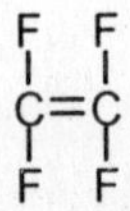

Fig. 1.16.   Fig. 1.17.

Substitution of all four atoms of Hydrogen by those of Fluorine yields monomer of Tetrafluoroethylene (Fig. 1.17).

Polymerisation is not limited to formation of macromolecules out of one type of monomers only. The process can be carried out with two or more monomers such as with Acrylonitrile and Styrene to form Acrylonitrile Styrene (SAN), a copolymer (Fig. 1.18).

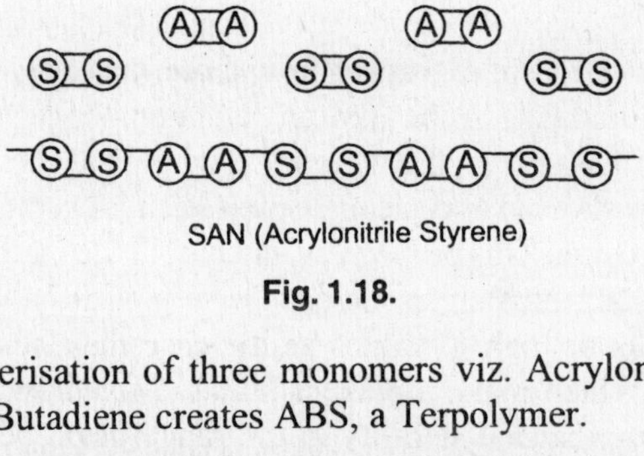

SAN (Acrylonitrile Styrene)

Fig. 1.18.

Polymerisation of three monomers viz. Acrylonitrile, Styrene and Butadiene creates ABS, a Terpolymer.

Besides Polymerisation, **Polycondensation** is another process of forming plastics out of two different types of building blocks. It may be remarked here that unlike polymerisation, this process sets a by-product free. Nylons (Polyamides) are formed by this process. Their names such as PA 6 6, PA 6 10, PA 6 12 are derived from the number of carbon atoms in each part of the combined monomer. For example, PA 6 10 is [$-NH (CH_2)$ 6NH$-$ $-CO(CH_2)_8CO-$].

If three or more groups are joined together by polycondensation, the final product is a thermosetting plastics. It is no more linear like a thermoplastics but has a three dimensional molecule.

Polyaddition is the process of forming a new product out of two different molecular building blocks without giving rise to a by-product. Polyurethanes are produced by polyaddition of Isocynate and Dialcohol. The process differs from polymerisation in that atoms are exchanged between the two monomers.

Plastics for Blow Moulding. As explained before, only thermoplastics can be used for production of articles through blow moulding process. In principle, all thermoplastics can be blow moulded but the most common ones for packaging are the Polyolefins viz. Low Density Polyethylene, High Density Polyethylene and Polypropylene as well as Polystyrene, Polyvinyl Chloride and Polyethylene Terepthalate. Acrylonitril Butadiene Styrene, Polyamides, Polyacetal, and Polycarbonate are being increasingly employed for blow moulding of technical components.

The very basis of extrusion blow moulding, viz. inflation of a hot thermoplastics pipe into an intended shape, makes one property imperative in the material. The pipe at the mouldable temperature should have adequate melt strength to retain its form during extrusion. This property is inherent in the material grades with very long molecular chains. These grades have low melt flow indices or high K-values as the case may be. The other attributes, which make a plastics suitable for blow moulding, are adequate thermal stability in the range of processing temperatures, good flowability, stretchability and weldability.

Common thermoplastics for blow moulding

A. Amorphous thermoplastics

1. Polystyrene (PS)

Also referred to as the General Purpose Polystyrene, it is an amorphous thermoplastics with a glossy surface. It is transparent in natural state. It is hard and brittle. The wear resistance is average. As it absorbs water only to a negligible extent, its dimensional stability is excellent.

- It burns with a sooty flame and smells sweet.
- It is resistant to weak acids and alkalis but swells in gasoline, benzene and other organic solvents.
- It is prone to stress cracking.
- It emits a metallic sound when dropped on hard surface.

Specific gravity	1.05
Glass temperature	80-100°C
Blow moulding temperature	180-210°C
Working temperature without load	
Short term	80-85°C
Long term	65-75°C
Mould shrinkage	0.6%

Uses

Packaging for food, medicines, cosmetics etc.

The following **copolymers** of Polystyrene overcome various deficiencies like brittleness of Polystyrene, retaining other attributes.

2. High impact polystyrene (HIP)

The General Purpose Polystyrene (G.P.P), modified with Butadiene, gains impact strength but loses its transparency and much of its lustre. The wear resistance is average. It absorbs some water and needs pre-drying before processing.

- It burns with a sooty flame.
- It is resistant to stress cracking.
- It is resistant to weak acids and alkalis. It is attacked by organic solvents.

Specific gravity	1.05
Glass temperature	80-90°C

Blow moulding temperature	180-210°C
Working temperature without load	
Short term	60-70°C
Long term	50-60°C
Mould shrinkage	0.6-0.8%

Uses

Packaging for food, medicines, cosmetics, toys etc

3. Acrylonitrile Styrene (SAN)

This copolymer of Styrene retains the lustre and transparency of the original material. It is tough and hard and has better heat resistance.

- It absorbs more water and needs pre-drying before processing.
- It burns like G.P.Polystyrene.
- It is resistant to weak acids and alkalis. It is dissolved by organic solvents.
- It is resistant to stress cracking.
- It has better heat resistance than Polystyrene.

Specific gravity	1.08
Glass temperature	106°C
Blow moulding temperature	180-230°C
Service temperature without load	
Short term	95°C
Long term	85°C
Mould shrinkage	0.6-0.8%

Uses

Packaging for food, medicines, cosmetics, etc.

4. Acrylonitril Butadiene Styrene (ABS)

ABS has better hardness, scratch resistance, toughness, heat resistance and resistance to chemicals than the other derivatives of Styrene.

- It is yellowish opaque in natural state. A special formulation is transparent.
- It absorbs more water and must be pre-dried before processing.
- It is resistant to stress cracking.
- It burns with a sooty flame.

Special attribute. ABS can be metalised.

Specific gravity	1.04-1.06
Glass temperature	85-100°C
Blow moulding temperature	200-215°C
Service temperature without load	
Short term	90-100°C
Long term	80-85°C
Mould shrinkage	0.6-0.8%

Uses

Packaging for cosmetics etc.

5. Cellulose Acetate (CA)

Cellulose Acetate as well as other derivatives of cellulose such as Cellulose Nitrate, Cellulose Butyrate and Cellulose Propionate are thermoplastics formed by modification of the natural cellulose.

- Cellulose Acetate is transparent and has high gloss.
- It possesses very good scratch resistance.
- It is tough and hard.
- It absorbs vibrations.
- It does not attract dust.
- It is resistant to UV-rays
- It is resistant to stress cracking.
- It is resistant to mineral oils, fats, benzene, weak sulphuric acid.
- It is not resistant to strong acids, alkalies, solvents on the

basis of ketones and esters and various chloro-carbo-hy-
drates.
- It absorbs water over 4%.

Specific gravity	1.3
Glass temperature	50-60°C
Blow Moulding temperature	165-195°C
Service temperature without load	
Short term	80°C
Long term	70°C
Mould shrinkage	0.6-0.8%

Uses

Containers for oil, paints, powders, lighter fuel, toys, lamp shades
etc.

6. Unplasticised Polyvinyl Chloride (uPVC)

Unplasticised Polyvinyl Chloride or rigid Polyvinyl Chloride is
one of the most widely used plastics. Basically, it is an unstable
polymer but a broad palette of stabilisers, additives and copolymers have made it the most prolific of synthetic materials so
that it offers formulations for extremely diverse applications.

- It is a transparent thermoplastics with high mechanical
  strength, stiffness and hardness.
- It is highly resistant to chemicals.
- It is a good insulator for low voltages.
- It is resistant to stress cracking.
- It has a narrow temperature range for processing.
- It can be used for outdoor applications.
- It burns with a yellowish-orange pungent flame emitting
  chlorine but is self-extinguishing.

Specific gravity	1.4
Glass temperature	80°C
Blow moulding temperature	175-180°C

Service temperature without load
    Short term   75°C
    Long term   65°C
Mould shrinkage   0.5-0.8%

Uses

Packaging for household liquids.

7. Polycarbonate (PC)

Polycarbonate is an engineering thermoplastics, which exhibits very high stiffness, strength and hardness between extreme temperatures of −150° C and +135° C. It is transparent like glass.

- It is a very good electrical insulator.
- It has very high heat resistance.
- It is prone to stress cracking.
- It is resistant to weak acids and alkalis.
- It burns with a sooty, pungent smell of phenol but is self-extinguishing.
- It absorbs water and necessitates pre-drying before processing.
- It can be used in contact with foodstuff.

Specific gravity   1.2
Glass temperature   150°C
Blow moulding temperature   250-275°C
Service temperature without load
    Short term   −135 to +160°C
    Long term   −135 to +135°C
Mould shrinkage   0.5-0.8%

Uses

Milk bottles, packaging for baby food, lamps etc.

8. Polyethylene Terephthalate (PET)

PET is linear thermoplastics Polyester of the semi-crystalline variety and as such it is white opaque. The share of the crystalline structure is about 30-40% but it can be reduced to such an extent with the help of co-monomers that the resultant polymer is transparent. Its remarkable properties are:

- Excellent hardness, stiffness
- Very good toughness, also at sub-zero temperatures.
- Good ageing characteristics. Low coefficient of friction.
- Good electrical insulation.
- Resistance to stress cracking.
- No reaction with water, dilute acids, alcohols, oils and fats etc. at room temperature.
- Reaction with high temperature steam, alkalis, oxidising acids, and organic solvents.
- Fire resistant, but burns with a sooty flame.
- Absorbs very little water but needs pre-drying.
- Can be used with foodstuff.

Note. PET is mostly processed through Injection stretch blow moulding.

(Data for amorphous PET)

Specific Gravity	1.33
Glass temperature	79°C
Injection moulding temperature	260-290°C
Stretch blow moulding temperature	90-110°C
Service Temperature without load	
Short term	180°C
Long term	100°C
Mould shrinkage	0.2-0.4%

Uses

Transparent bottles for aerated beverages, vegetable oils, etc.

9. Glass Polymers

Glass polymers are copolyesters and are predominantly amorphous. Their outstanding properties are:

- Superior chemical resistance
- Resistance to stress cracking
- Resistant to oils, fats, alcohols, aliphatic hydrocarbons.
- Gloss better than that of PMMA and SAN
- Water clear like PMMA
- High scratch resistance
- High resistance to breakage; virtually shatterproof
- Good dimensional accuracy due to low mould shrinkage
- Rigid but tough even with thin walls
- Clip and snap fit possible
- Low crystallinity even in thick parts; clarity unimpaired.
- Barrier properties to Oxygen and moisture with thick wall

Note. As Glass polymers absorb moisture, they must be pre-dried for 6 hours at 65° C.

Specific Gravity	1.18-1.37
Glass temperature	80-84°C
Blow Moulding temperature	200-240°C
Mould shrinkage	0.3-0.6%

Uses
Fancy packaging for cosmetics.

10. Polysulphone (PSU)

Polysulphone is a high temperature, engineering polymer with properties, which enable it to replace light metal alloys in many applications.

- It possesses very high hardness, stiffness and tensile strength
- It can withstand very high temperatures.
- It has very good electrical properties, also at higher temperatures and under high humidity..
- It can withstand acids, alkalies, benzene, oils, fats, detergents and salt solutions.

- It has good hydrolyses stability.
- It is attacked by ketones, polar organic solvents, aromatic and chlorinated carbohydrates.
- It can be used with foodstuff.
- Its water absorption is very low.
- It tends to stress cracking with certain solvents.
- It is difficult to ignite.
- It is transparent with light yellow tinge.
- It has low notch strength, which can be improved through blending.

Special attribute. PSU can be galvanised.

Specific gravity	1.24
Glass temperature	190°C
Blow moulding temperature	275-300°C
Service temperature without load	
Short term	200°C
Long term	150-170°C
Mould shrinkage	0.7%

Uses

Automobile components under the hood, technical components for high temperature applications.

B. Semi-crystalline thermoplastics

Polyethylenes of different densities along with the Polypropylene constitute the group of polyolefines. The common attribute of these semi-crystalline thermoplastics is their specific gravity, which is less than one so that they all float in water. The monomer Ethylene is the simplest one, having two atoms of carbon and four of hydrogen (C_2H_4), which is also the basic building block of wax.

There are two distinct forms of Polyethylene: the Low Density and the High Density Polyethylene. Most of their properties are similar; the difference lies in the extent.

1. Low Density Polyethylene (LDPE)

LDPE is translucent to whitish-opaque, depending upon the wall thickness. It is flexible.

Its molecules can arrange themselves parallel to one another in crystalline structures among the mass of randomly scattered molecules while cooling from molten state to the solid phase. The degree of crystallisation may reach 40-50% depending upon the rate of cooling and the linearity of the molecules. The crystalline zones contribute stiffness.

- It possesses very high toughness.
- It has excellent electrical insulation properties.
- It absorbs very little water.
- It has very good resistance to all chemicals at room temperature.
- L.D.Polyethylenes with higher molecular weights are more resistant to stress cracking.
- It burns without soot and smells like wax.

Specific gravity	0.91-0.93
Melting temperature	105-110°C
Blow moulding temperature	140-170°C
Service temperature without load	
Short term	80-90°C
Long term	60-70°C
Mould shrinkage	1.5-3%

Note. The cross-linking varieties of LDPE, which become permanently set after treatment, have higher working temperatures.

Uses

Packaging for corrosive chemicals, liners, flexible bottles, toys etc.

2. High Density Polyethylene (HDPE)

HDPE has a molecular structure, which is more linear than that of LDPE. It is also stiffer than the LDPE.

- In natural state, HDPE is whitish in colour. Very thin layers may appear translucent.
- HDPE has poor scratch resistance. Ultra High Molecular High Density versions possess very good surface hardness and sliding properties.
- It is resistant to almost all chemicals, except to strong acids, at room temperature.
- It is prone to stress cracking.
- It is an excellent insulator to electricity.
- It burns with a bright, sootless flame, emitting smell of wax.

Specific gravity	0.94-0.97
Melting temperature	130-150°C
Blow moulding temperature	160-190°C
Service temperature without load	
Short term	90-100°C
Long term	70-80°C
Mould shrinkage	1.5-5%

Uses

Containers for food and general merchandise, petrol tanks and storage tanks for oils, containers for transport, bottles, toys etc.

3. Polypropylene (PP)

Polypropylene is the lightest member of the polyolefin family.

- It has high toughness, hardness and the tensile strength.
- It has high heat resistance. It can withstand boiling water.
- It has very good chemical resistance.
- It is translucent to opaque; some special formulations are transparent in thin sections.
- It burns with a soot-free flame and smells like wax.
- It becomes brittle at sub-zero temperatures.
- It is sensitive to UV-rays.

Note. With the help of nucleating agents, the pattern of crystallinity can be altered, in that instead of a few big clusters of crystals, many smaller clusters are formed. This structure results in less dispersion of light, conferring on the material more transparency. The nucleated polypropylene shrinks more but also more uniformly. The warpage is also reduced.

Special attribute. Thin, integral hinge out of PP is practically indestructible.

Specific gravity	0.88-0.91
Melting temperature	158-165°C
Blow moulding temperature	190-230°C
Service temperature without load	
Short term	140°C
Long term	100°C
Mould shrinkage	1.5-2.5%

Uses

Bottles for household fluids and detergents, canisters and drums for chemicals, jerry cans, containers for medical infusion solutions, technical components for automobiles and household gadgets, absorber plates for solar cells, sports goods, toys etc.

4. Polyamides (PA)

The specific aim of development of **Polyamides** was the artificial yarn. Patented in 1937, PA 66 was introduced in USA in 1940 in the form of Nylon stockings.

In Germany, simultaneous research led to the development of PA 6 and PA 66 for injection moulding and extrusion. They were found to be eminently suitable for engineering applications because of the following properties.

• Very high strength, stiffness and hardness.
• Very good heat endurance.
• Excellent wear resistance, good sliding characteristics, low friction.

- Very good shock absorption and damping.
- Resistance to solvents, gasoline, oils and fats.

Special attribute. All polyamides absorb moisture, which influences their properties. The toughness increases considerably but dimensions also increase and the electrical insulation decreases.

Attributes	PA 6	PA 6 6	PA 6 10	PA 11	PA 12
Specific gravity	1.13	1.14	1.08	1.04	1.02
Melting temperature°C	220	255	215	185	180
Service temp. without load					
Short term °C	140-180	170-200	140-180	140-150	140-150
Long term °C	80-100	80-120	80-110	70-80	70-80
Mould shrinkage %	0.8-2.1	1.0-2.2	0.5-2.8	0.5-1.5	0.4-0.6
Moisture intake %	2.5-3.5	2.5-3.1	1.2-1.6	0.8-1.2	0.7-1.1
Processing temperature °C	250-260	270-290	230-250	200-230	200-230

Uses

Technical components for automobiles and household machines.

Special note. Polyamides are semi-crystalline and as such opaque but the aromatic polyamides are transparent.

The number given after the abbreviation PA denotes the number of carbon atoms in the monomer (PA 6, PA 11, PA 12). The double number indicates a monomer with two molecules and stands for the number of carbon atoms in each molecule (PA 6 6, PA 6 10).

5. Polyacetal (POM=Polyoxylmethylene)

Polyacetal is an engineering thermoplastics from the semi-crystalline family. Its distinguishing properties, as listed below, make it an ideal material for technical applications.

- High hardness and stiffness.
- High toughness, also upto –40°C
- High heat resistance
- Low water absorption, good dimensional stability
- Good electrical properties
- Resistant to petroleum, oils, fats, alcohols, weak acids and alkalis, detergents.
- Excellent surface hardness and wear resistance
- Very low friction, self-lubricating behaviour
- Polyacetal is whitish in natural state. It burns with a pungent bluish flame smelling of formaldehyde.

Special attribute. Polyacetal is suitable for long lasting integral hinges.

Specific gravity	1.42
Melting temperature	175°C
Blow moulding temperature	170-190°C
Service temperature without load	
Short term	110-140°C
Long term	90-110°C
Mould shrinkage	1.0-3.0%

Uses

Floats for fuel tanks, gas ampoules, aerosol containers

Plastics are generally referred to under their acronyms. The following list gives acronyms of the most common polymers.

Acronyms for some Common Polymers.

ABS	Acrylonitrile Butadiene Styrene
CA	Cellulose Acetate
CAB	Cellulose Acetate Butyrate
CAP	Cellulose Acetate Propionate
CN	Cellulose Nitrate

EVA	Ethylene Vinyl Acetate
FEP	Fluorinated Ethylene Propylene
HDPE	High Density Polyethylene
HDHMWPE	High Density High Molecular Weight Polyethylene
HIPS	High Impact Polystyrene
LCP	Liquid Crystal Polymer
LDPE	Low Density Polyethylene
LLDPE	Linear Low Density Polyethylene
MF	Melamine Formaldehyde
PA	Polyamide
PAI	Polyamide Imide
PAN	Polyacryle Nitrile
PBTP	Polybutylene Terephthalate
PC	Polycarbonate
PEEK	Polyether Ether Ketone
PEI	Polyetherimide
PES	Polyether Sulphone
PET/PETP	Polyethylene Terephthalate
PF	Phonol Formaldehyde
PI	Polyimide
PMMA	Polymethyl Metharcylate
POM	Polyoxymethylene
PP	Polypropylene
PPO	Polyphenylene Oxide
PPS	Polyphenylene Sulphide
PS	Polystyrene
PSU	Polysulphone

PTFE	Poly Tetra Fluoro Ethylene
PU/PUR	Polyurethane
PVC	Polyvinyl Chloride
PVDC	Polyvinylidene Chloride
PVDF	Polyvinylidene Fluoride
SAN	Styrene Acrylonitrile
SB	Styrene Butadiene (HIPS)
TPE	Thermoplastic Elastomer
TPU	Thermoplastic Polyurethane
UHMWHDPE	Ultra High Molecular Weight High Density Polyethylene
UF	Urea Formaldehyde

Blow Moulding Process, Equipment and the Mould

Extrusion blow moulding is one of the major processes for production of hollow articles with openings smaller than the body, out of thermoplastics.

In simple terms, blow moulding is like blowing up a balloon in a confined space. Blow moulding as an industrial production process consists of inflating a hot, deformable thermoplastics pipe in an enclosed cavity till it assumes the shape of the said cavity.

The above definition implies:

- Production of a thermoplastics pipe.
- Its transportation to a split mould with a cavity of the required shape.
- Arrangement to open and close the mould to introduce the pipe and to take out the blown object after inflation and cooling.
- Arrangement of inflation of the pipe.

It follows that a blow-moulding machine would consist of the following units:

- Pipe producing unit, or the extruder

- Unit to hold and move the split mould, or the mould closing unit
- Inflating device, or the blowing unit

The mould is a separate entity, which helps to produce the desired hollow object on a blow-moulding machine.

1. The Extruder (Fig.2.1)

A typical blow moulding extruder consists of a hollow steel barrel with a precise cylindrical bore, in which a closely fitting screw is made to rotate with the help of an electrical or a hydraulic motor. The barrel is equipped on its outside with a number of electrical band heaters, and in most cases with cooling spirals too, to achieve a temperature within a closely controlled range. A hopper, fitted over an opening at one end of the barrel, feeds in the plastics raw material, which is conveyed forward, compressed, plasticised and metered out at the other end by the rotating screw. The heat needed for melting or "plasticising" of the material comes partly from the electrical heaters around the barrel but mainly from the friction or "shear" due to rubbing of the plastics material against the inside wall of the barrel and of

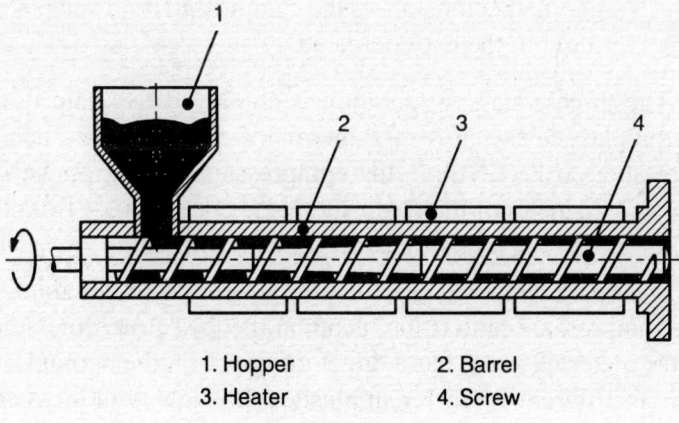

| 1. Hopper | 2. Barrel |
| 3. Heater | 4. Screw |

Fig. 2.1.

different layers of the melt against one another. In order to fulfil its manifold functions, the screw comprises three zones (Fig. 2.2):

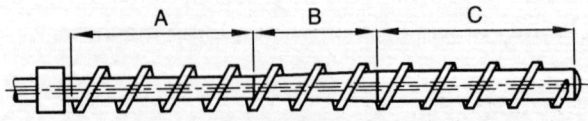

A. Feeding zone
B. Compression zone
C. Metering zone

Fig. 2.2.

    I. The Feed Zone, where the screw has maximum flute depth, is meant to receive the material from the hopper and to convey it to:

   II. The Transition or the Compression Zone with progressively decreasing flute depth, which compresses the resin, plasticises it to a void free viscous fluid and transfers it to :

  III. The Metering Zone of the screw having uniform flute depth. It homogenises the compressed fluid and meters it out to the extruder head.

The three zone configuration is employed for majority of thermoplastics except for the thermally sensitive ones such as unplasticised PVC. Here, the compression is progressive i.e. the core diameter of the screw increases continuously from the feed zone till the metering zone in linear fashion.

It may be remarked here that the degree of compression or the compression ratio (Flute depth at the feed zone/flute depth in the metering zone) has a direct influence on the output. This ratio is different for different plastics. It is low with thermally sensitive plastics like PVC—less than 2:1— but higher with those with a broad temperature range. Screws for processing of Polyolefins have a compression ratio of 4:1.

The barrel is provided with a cooling circuit in the feed zone to prevent "bridging" or building of lumps of the raw material through the heat coming from the adjacent zone. The lumps block transport of material to the next section.

The output end of the extruder is equipped with a cross head that changes the direction of the flow of the melt from horizontal to vertically downwards and shapes it into a pipe by means of a built-in torpedo (Fig. 2.3). Provision is made to fit dies and cores of various sizes to produce pipes of different diameters and thickness as required for the product to be blow moulded. The plastics pipe extruded out at this end is generally referred to as a "parison".

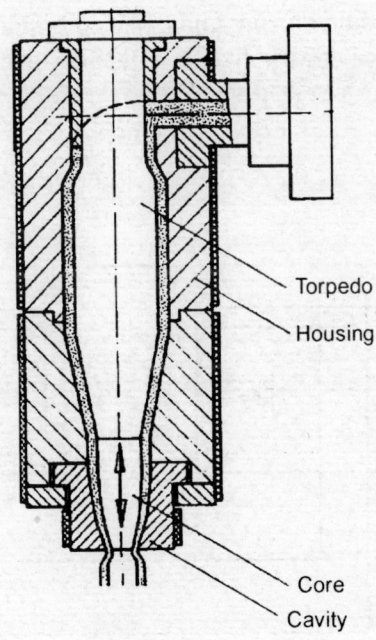

Torpedo

Housing

Core

Cavity

Fig. 2.3.

An air hole along the axis of the torpedo and having con-
nection outside the cross head through one of its spider legs
(not shown here) serves to supply support air to prevent collapsing of the parison during cutting and transport to the mould.

2. Mould Closing Unit

Arranged directly below or besides the cross head is a mould
closing unit, consisting of two vertically mounted platen upon
which the two halves of the mould are fixed (Fig.2.4). The platen
are made to move to and fro hydraulically or mechanically, to
close or open the mould. The last part of the closing movement
is slowed down to avoid hammering of the mould halves as
well as to enable the pinch off part of the mould to push up the
material in the cavity and thus thicken the welding seam. (See
also chapter VII, the Pinch off). The platen may have tie bars
for guidance or they may be guided from behind.

The mould closing units, placed besides the cross head and
not directly below it, have two tie bars placed diagonally so that

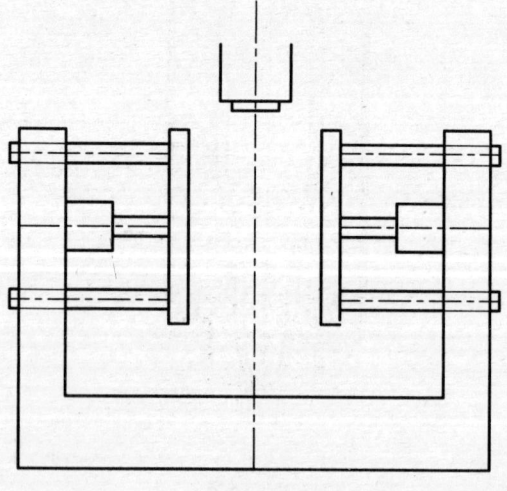

Fig. 2.4.

these do not come in the way of the hanging parison as the closing unit travels to the cross head. There is no such restriction
for the closing units situated below the cross head, receiving
parison from above. The locking force has to compress and severe the flash during closing of the mould and then has to withstand the blowing pressure exerted by compressed air during
the phase when the parison is inflated. Some manufacturers provide additional cylinders besides those for the closing movement in order to provide hammering strokes on closed moulds.
The action decreases the thickness of material caught between
the cutting edges of the mould and ensures an easy and clean
deflashing.

There are several ways of bringing the parison to the mould.
Either a pair of grippers severs the parison under the cross head
and transports it to the mould closing unit or the mould closing
unit itself may travel to the cross head, close the mould over the
parison and return with it to its initial site where blowing takes
place. With some thermoplastics, it is necessary to incorporate
additional devices like hot or cold knives to cut the parison before transportation.

7. The Blowing Unit (Fig. 2.5)

An interchangeable blowing spigot, pin or mandrel, capable of
moving up and down i.e. into the mould or out of it and the
arrangement to hold and move it and to send compressed air
through it is a device, which forms the third essential part of a
blow-moulding machine. Although the blowing mandrel is a
part of the blowing unit of the machine, it has a direct relationship with the mould and hence it must be made for and matched
with each mould. Its primary function is to introduce compressed
air into the parison to inflate it. The blowing station also performs the function of ejection by stripping the blown article off
the mandrel by means of a stripper plate or a sleeve placed around
the mandrel.

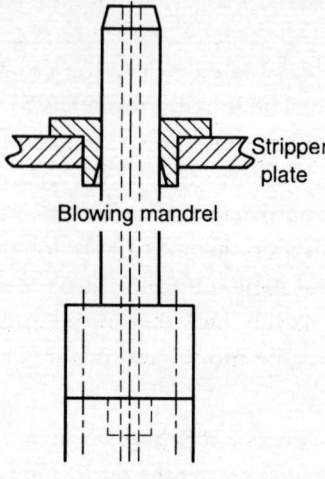

Blowing mandrel

Stripper plate

Fig. 2.5.

The Process

A suitable length of the parison is cut and brought between the mould halves, which close over it and seal its open ends (Fig. 2.6). The blowing mandrel, introduced in the mould generally through the neck or mouth of the intended hollow article such as a bottle, before or after closing of the mould depending upon the machine system, lets in compressed air which inflates the parison till it touches the walls of the cavity in the mould and can expand no more (Fig. 2.7). The mould, cooled by circulation of water in channels provided in it for this purpose, dissipates the heat of the blown parison till it is cold and rigid enough to be ejected without deformation. At this stage, the mould-closing unit opens to move apart the two mould halves. The blow-moulded product, which sticks onto the blowing mandrel, can be stripped off and the next cycle can begin.

Blow moulding is, as a rule, a continuos process. The output of the extruder is adjusted in tune with the moulding parameters so that the next parison in proper length is just ready when the mould has become free to receive it.

Although the foregoing description pertains to a blow-mould-

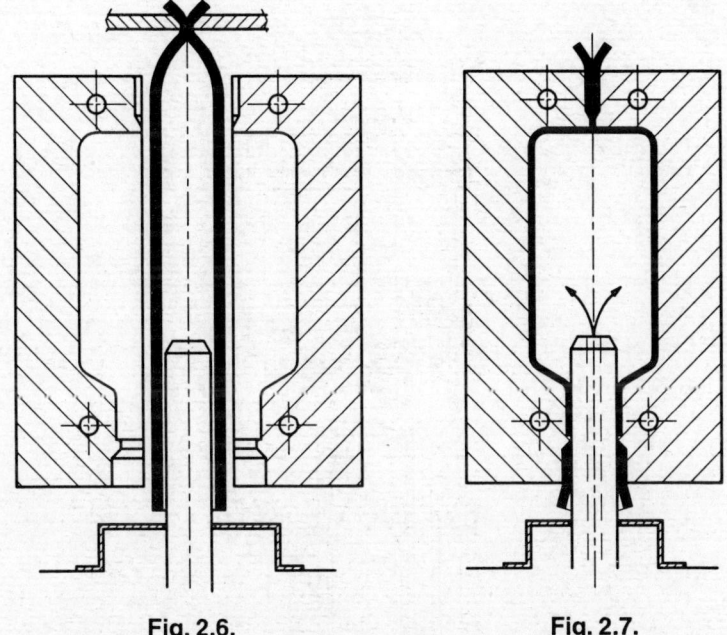

Fig. 2.6.   **Fig. 2.7.**

ing machine with a horizontal, single screw extruder, bringing out the parison vertically downwards, it may be pointed out that a number of variations are possible in this principle. The extruder may be placed horizontally, at an angle or vertically. In the latter case, no cross head is required to change the direction of the parison. The cross head may be designed to extrude one or more parisons for multi-cavity production.

For large articles like drums, storage containers and tanks etc., requiring parisons weighing many kilograms, the process of continuous extrusion as dealt with so far, suffers from many drawbacks. The inordinately long time for extrusion is bound to introduce temperature variations in a long parison. The heavy parisons tends to neck i.e. becoming thinner at the top end near the crosshead. The method adopted to circumvent these shortcomings is to introduce an accumulator head at the end of the extruder, instead of the crosshead (Fig. 2.8). The extruder feeds the melt in the storage cylinder with the piston simultaneously

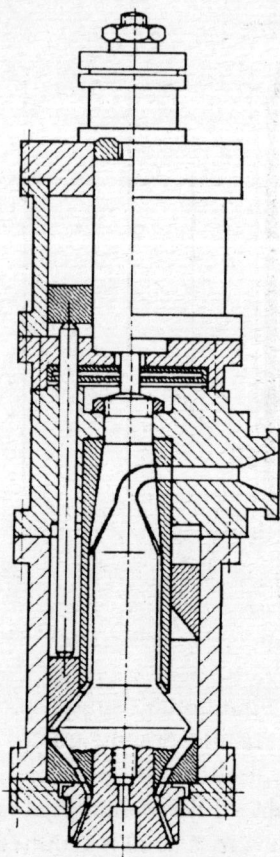

Fig. 2.8.

going up till the predetermined amount of the stock has been fed in. The extruder has to stop at this juncture. The piston comes in action and pushes the melt in a fast stroke out of the crosshead as a parison. The process is no more continuous but cyclic.

There are many variations in the accumulator head design as also in those of the crosshead but they have no influence on the design of the mould. This is why, it is not intended to enumerate and describe the various designs developed by numerous machine manufacturers.

The Blow Mould

A blow mould is required to perform several functions which may be summed up as:

- To transform the parison into the shape of a hollow article
- To seal the open ends of the parison
- To cool the blown article
- To separate and cool the superfluous material

Fig. 3.1 depicts a blow mould with all usual components and design features. It will be observed that the mould consists of two halves, almost identical in construction. The shape of the cavity in each half, however, solely depends upon that of the product. For articles with unsymmetrical shapes, the two halves can differ substantially from each other.

A simple blow mould consists of the following components:

1. Back plates
2. Cavity blocks
3. Guiding elements
4. Base inserts
5. Neck inserts
6. Cutting rings
7. Top inserts

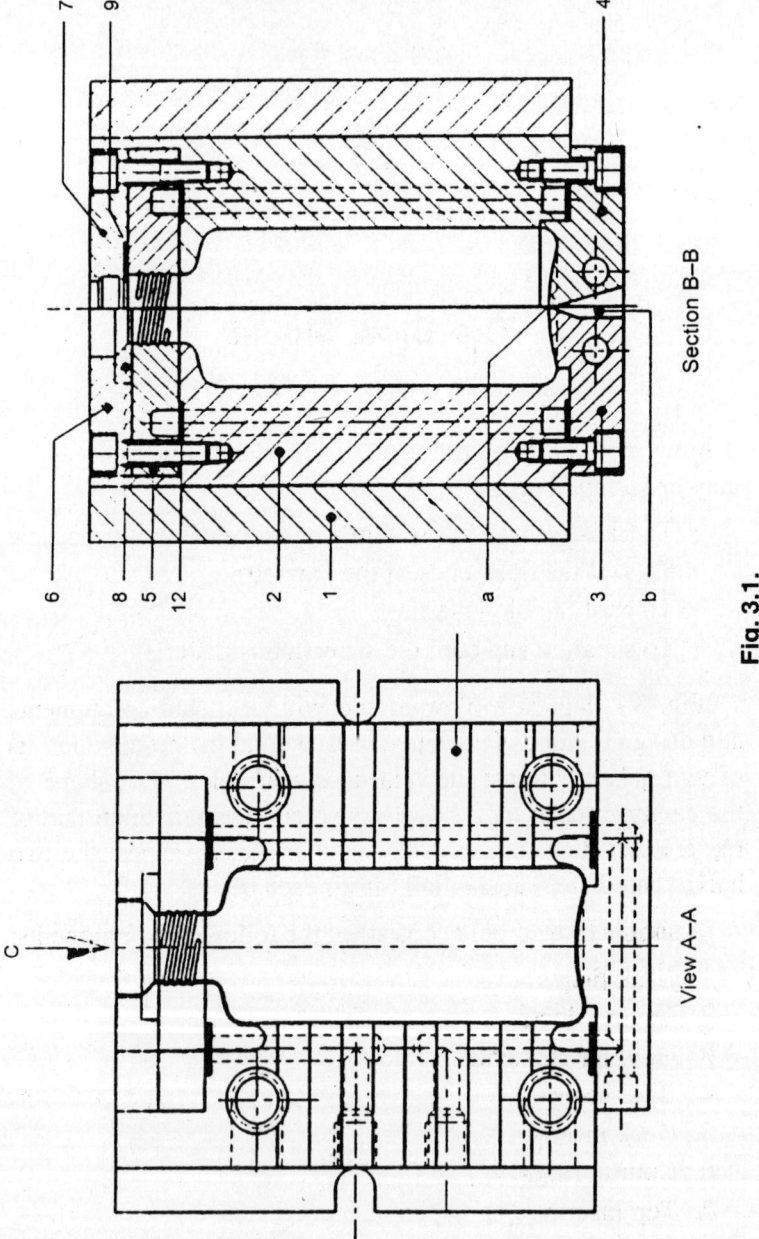

Fig. 3.1.

View A–A

Section B–B

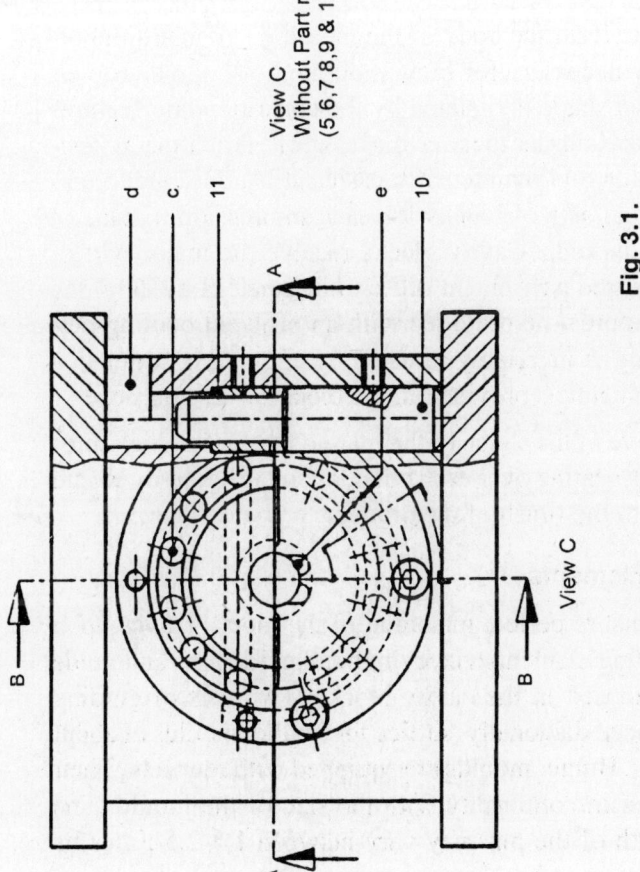

Fig. 3.1.

1. Back plates

The back plates (Part no. 1) are used to mount the mould halves on the machine platen by means of clamps or directly by screws. For this purpose, they are made to project on the sides. They may also serve to interconnect cooling lines, particularly in multi-cavity moulds. In order to clamp small moulds having very little weight, the back plates may conveniently be replaced by flat bars, about 30-40 mm broad, housed in milled slots at the back of the cavity blocks.

2. Cavity Blocks

Cavity blocks form the body of the mould and contain important features, necessary for blow moulding of hollow articles. Their size and shape is dictated by the corresponding features of the product. In order to make the tasks of fabrication, adjustment, matching and maintenance easier, it is advisable to employ individual sets of blocks for each impression in case of multi-cavity moulds. Cavity blocks receive the major part of the heat from the parison. In order to dissipate it quickly, the cavity blocks must be provided with an efficient cooling network. Cooling is invariably effected by circulating water in a labyrinth of channels provided in the block for this purpose.

Irrespective of the shape of the mating faces, the block must have a perfect seating over each other, as any irregularity would be reflected in the finish of the product.

3. Guiding Elements

In order to ensure perfect matching of the mould halves upon closing, guiding elements viz. cylindrical guide pins and guide bushes are housed in the cavity blocks. Two sets of guiding elements, placed diagonally, suffice for smaller moulds of about 20 cm. height. Bigger moulds are equipped with four sets. Their size is chosen in conformity with the size of the mould. The guiding length of the pin may vary between 1,5-2,5 times its diameter.

The guiding elements should be located sufficiently away from the cavity impression to avoid entanglement of the guide pin with the parison. Unless the melt caught accidentally in the bush has an outlet, perfect closing of the mould will be prevented. In extreme cases, especially with cast moulds, the entrapped plastics compressed under the bush, may generate cracks in the cavity block. A material escape slot directly below the guide bush banishes the danger.

Guiding elements are hardened and ground to a sliding fit.

4. Base Inserts

As the name suggests, this component shapes the base of the article. Base constitutes a critical area due to following reasons. The parison is sealed and the flash or the superfluous material is separated here by means of cutting edges. Part 3 and 4 show two different designs of this feature.

The flash consists of dense mass of hot plastics and contains a large amount of heat. Unless cooled intensively, it will stick to the ejected article. The base inserts are, therefore provided with individual cooling circuits, except in very small moulds. In addition, the flash section may be furrowed in order to increase the surface area for more intensive cooling.

The base inserts have to perform the task of sealing and cutting, which leads to wear. This is why; they are fabricated out of steel and hardened. In spite of that, they must be reground a few times during the course of their life. This also explains, why they cannot be made integral with the cavity block. An inserted base also acts as a venting path for the air trapped in the corners.

5. Neck inserts

Neck is another section, which underlies almost the same conditions as the base. Here too, the parison is welded and the flash separated. Unlike at the base, here the parison is subjected to little stretching so that there is concentration of heat in the neck area. The flash is also another reservoir of heat, which must be dissipated quickly. Neck insert is always designed as a replaceable component. The accuracy required in the threads of the neck also supports this step. Here too, the neck insert fulfils the secondary function of venting. It is fabricated out of steel and hardened. Unless the mould is very small, the neck inserts possess their own cooling circuit.

6. Cutting rings

The top edge of the neck is formed by cutting rings (half rings to be precise), two variations of which (Parts 8 and 9) have been shown in the mould design. Part no. 8 is fashioned for the "Plunge-in" type of the blowing mandrel. The inside diameter of the ring, corresponding to the outside diameter of the blow moulded article, is flared out like a funnel after a straight length of about 1 mm. The included angle may vary from 60 to 90 degrees.

A thin cutting ring (Part no. 9) is used with straight calibrating mandrels, which enter the parison before closing of the mould. Its inside diameter equals that of the blowing mandrel, which in turn is usually the internal dimension of the neck of the article being blown.

In lieu of the function the cutting rings are called upon to perform, it is essential that they be hardened. They are subjected to a maximum amount of wear. They should, therefore, be always designed as separate replaceable units.

7. Top inserts

Primarily meant to hold the cutting rings in place, the top inserts may also serve to seal the milled cooling slots in the neck inserts. The top insert (Part no. 7) also guides the blowing mandrel. A deep groove cut in it, serves as accommodation and escape for the flash, unavoidable at the ends.

Considering their function, the top inserts are fabricated out of steel and hardened.

Other Design Features

"a" denotes the **welding or the cutting edge**, incorporated generally in the neck and the bottom region to seal the parison and to separate the superfluous material from the blown article.

"b" depicts two common designs of the **flash compression zone**.

"c" represents cooling arrangement and is an essential feature of every blow mould. In this particular case, the cooling network has been formed by drilled holes arranged around the cavity and interconnected.

"d" shows an **escape channel** for material caught accidentally in the guide bush. It is housed in the cavity block below the guide bush.

"e" depicts the **flash escape slots** in the top insert used with the straight calibrating mandrels. Slots are milled in such a fashion that three or more projections are left to guide the mandrel.

"f" denotes the **venting** device for the air contained in the hollow cavity, consisting in this case of shallow slots, machined on the mating face of one of the mould halves.

It must be pointed out in conclusion that many variations of the features cited above are employed in different cases. The short description should serve as an introduction to the design of blow moulds, providing the reader basic knowledge of the subject and an understanding of the interrelationship of various mould components and features. Each aspect will be discussed in detail in the subsequent chapters.

Methods of Blowing

The variety of blowing systems prevalent in practice makes it imperative for a mould designer to study their principle and mode of functioning closely as they also influence the design of the corresponding blow mould. The more common types will be described here in detail, illustrating their relationship with and influence upon the design of the blow mould.

a. Blowing through the Cross Head

The straight core forming the inside of the parison and projecting out of the crosshead performs also the function of a blowing mandrel. The mould halves, with the neck of the cavity pointing upwards, close over the mandrel in a horizontal movement right below the cross head. The inside of the neck is formed by the mandrel. Fig. 4.1 depicts the process in all vital details. Although the method guarantees optimum uniformity of the neck of the blown object, it can understandably be employed for intermittent production only.

This mode of blowing puts a severe limit on the size of the parison as its inside diameter has to tally with that of the inner bore of the neck of the product. Depending on the dimensions of the body of the product, the stretching ratio may prove to be unfavourable and some sections of the article may turn out to

Compressed Air

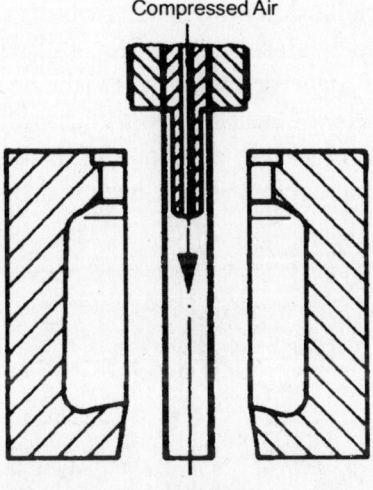

Fig. 4.1.

be too thin. If the parison were made thicker to compensate the drawback, the neck would have flash all around, which must be cleaned in a post-operation. The blowing air acts as a coolant and disturbs the heat balance of the crosshead.

The system of blowing through the cross head is known as the Plax method after the inventors of the blow moulding process but is almost obsolete now. It is sometimes used in hand moulding machines.

b. Blowing from Below

Developed by KAUTEX of Germany in early fifties, the system employs a vertical blowing mandrel situated in line with the parison far below it but otherwise independent of it. As soon as the required length of the parison is extruded, a pair of grippers holds it below the crosshead, severs it and brings it onto the mandrel. Now the mould halves close over the parison and the mandrel, squeezing the parison over the later so that the inside of the article neck is formed by the mandrel and the superfluous material around the neck, shoulders and the base is squeezed out into the flash pockets (Fig. 4.2). The mould halves contain thin half rings for cutting where the neck contour ends. The inside diameter of these rings matches with the outer diameter of the mandrel so that upon closing of the mould halves, the superfluous parison beyond the neck is sheared off (See also chapter III).

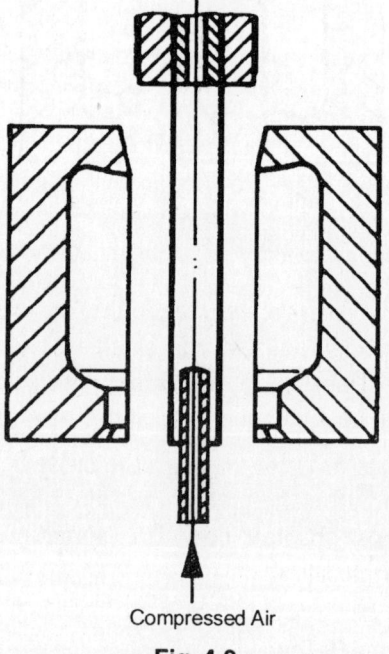

Compressed Air

Fig. 4.2.

The mould opens upon conclusion of the cycle. The blown article holds onto the mandrel. Ejection is effected by means of a stripper plate around the mandrel. A downward stroke of the mandrel results in ejection of the article. A jet of air from the side blows away the stripped article clear of the mould.

The system requires the mould cavity pointing downwards. In other words, the product sits in the mould with its bottom up.

A distinct advantage of the system lies in the fact that with suitable modification, the blowing mandrel can also form a part of the parison-stretching fixture for products having rectangular or oval shape, such as a jerry can (Fig. 4.3). In short, the fixture has a dummy mandrel placed at the side of the mandrel. After the parison has been brought over them, they are pulled apart horizontally, the mandrel to its foreseen position viz. the neck and the dummy nozzle outside the mould, thus stretching the parison to the breadth of the article to be moulded. The independence of the blowing mandrel from the crosshead enables incorporation of air flushing and water-cooling arrangements in it.

Fig. 4.3.

Blowing from below shows its handicap if PVC is to be processed. The part of the parison, which will form the base of the product, is the one close to the crosshead. Its own weight exerts a pull on the parison and makes it thinner where it would be inflated the most. In case of PVC, it is not possible to employ a parison-programming device to increase the thickness locally. However, with unusual configurations where the base is smaller than the body of the article, blowing from below is as good for PVC as for other thermoplastics.

c. Blowing from Above

Some blow moulding machines have the blowing station located besides the crosshead and not below it as in the previous case. The parison is transferred to the mould either by means of grippers which cut it below the cross head and bring it to the mould at the blowing station or the mould platen assembly with the open mould halves travels to the crosshead, closes over the parison and returns with it back to its initial position viz. the blowing station. The blowing mandrels are not a part of the mould-closing unit. The blowing mandrel enters or "plunges in" the closed mould having the parison, from above. It is understood that the opening or the neck of the article in the mould points upwards (Fig. 4.4).

The plunging-in type of mandrel has two diameters. The front portion forms the inside diameter of the neck and the second diameter, which is bigger, fashions the top face of the article orifice. As the Fig. 4.5 illustrates, the edge of the second diameter shears off the superfluous part of the parison. A positive side effect of the plunging action of the mandrel is that plastics material is pushed in the mould, making the neck more solid.

The method, developed by BEKUM of Germany, is extensively employed in double station blow moulding machines. The parison is brought to the two moulds situated on the right and left side of the crosshead alternately or in the other version, the moulds travel to the crosshead by turns. The process is continuous and free of the drawback of a thinned parison for the base part. As obvious, the top part of the mould for a plunging-

Compressed Air

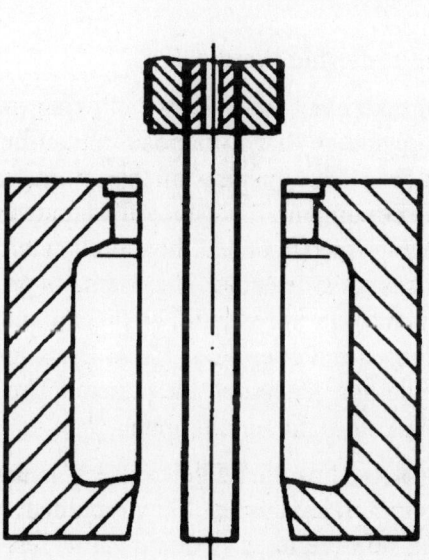

Fig. 4.4.

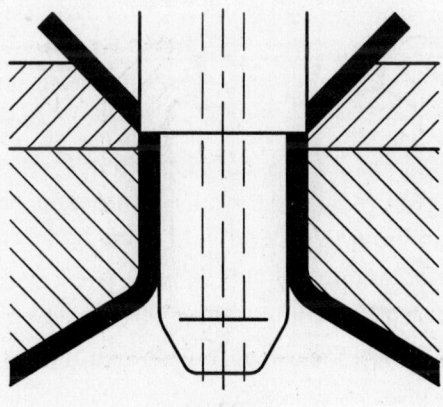

Fig. 4.5.

in mandrel would differ in design from that for calibrating type of blowing mandrel (See also Fig. 3.1, part 6 + 7).

d. Blowing with a Hypodermic Needle

A hollow needle can take over the role of a blowing mandrel with the additional advantage that it its position can be independent of the location of the opening of the article and the blowing station of the equipment. It is generally situated on the mould parting line, being fixed to one of the mould halves (Fig. 4.6). Other variations are also feasible. The needle is provided with forward movement with a hydraulic or pneumatic cylinder after the mould has closed over the parison. Blowing commences after the needle has penetrated the parison. The cylinder retracts the needle before the mould opens.

The method, also termed as the Miller process, is used extensively in the wide mouth containers, blow moulded with an extension of the neck, referred to as a "dome" or a "lost head" which is cut off in a post-operation. The blowing is effected by

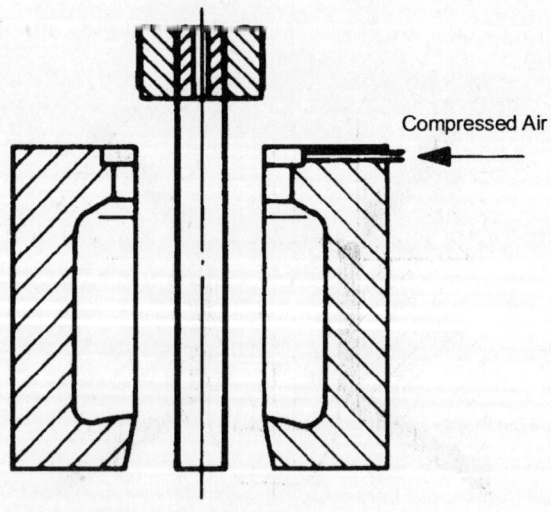

Compressed Air

Fig. 4.6.

piercing of the dome, whose height is taken as 1/3 to ½ of the head diameter. A unique application of this method of blowing is manufacturing of two identical containers, placed neck to neck, or a container and its lid, out of a single parison (Fig. 4.7).

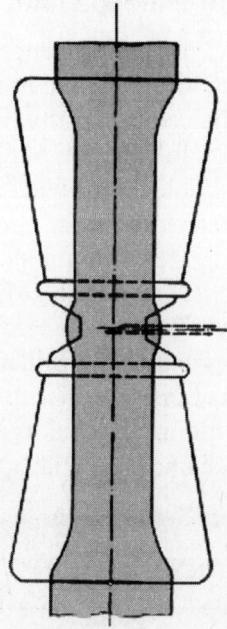

Fig. 4.7.

To make sure that the needle is not deflected by the parison, the former should be well guided between the two halves of the mould. The parison too should be suitably dimensioned and preferably preblown to facilitate entry of the needle which is usually cut at an angle in the front to give it a sharp edge.

Whereas by other methods of blowing, the blown article remains sticking to the blowing mandrel upon opening of the mould and can be stripped off, a blowing needle cannot perform this function. To this end, either the mould has to be equipped with special ejection devices such as an ejector pin actuated by springs or a pneumatic cylinder or a conventional mandrel has to be employed in conjunction with the blowing needle. The role of the mandrel is confined to preblowing the parison and holding on the blown article for stripping.

e. Blow Moulding with Entrapped Air

If a preblown parison, sealed at both ends, is squeezed between two halves of a blow mould with one or more cavities, the imprisoned air inflates the section of the parison covered by the hollow of the cavity. On full closure of the mould, the welding edges seal the blown article all around. The outcome is an article, blow moulded but without an aperture. The method is generally employed to mould a number of small articles like floats, bottles, ampoules, vials, balls, toys etc. out of a single parison (Fig. 4.8). It enables not only a more efficient utilisation of the machine capacity but also production of hollow articles where an aperture is inadmissible. The disadvantage if any, lies in the fact that the welding seam is present all around. The moulded articles are punched out with a press tool.

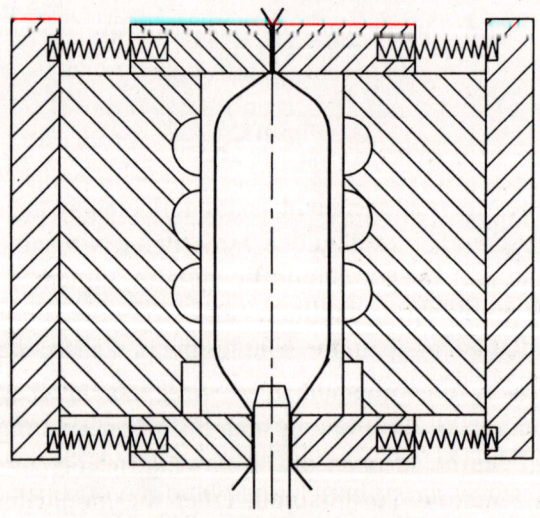

Fig. 4.8.

The Blowing Medium

As mentioned before, compressed air is employed as the means of inflation. The blowing usually takes place in two steps. The parison is inflated initially with compressed air at lower pressure of 3-4 bar, till it has touched the walls but not the corners. The low pressure has twofold action. It helps the parison to slip along the walls and it affords more time to the air in the mould to escape. Now the pressure is raised to 7-10 bar to stretch the parison to form the corners and to press against the cavity to assume its texture. The air is also helpful in cooling the product from inside if it is continuously renewed (see also Chapter 11).

The compressed air should be clean, dry as well as free of oil and dust particles. Normally, it is used at room temperature but some reduction in the cooling time is possible with air at lower temperatures. Too low a temperature of air, however, can cause condensation on the components in contact with it. Similarly, higher pressures too help accelerate cooling by pressing the hot article against the cold mould more intensely but call for higher mould locking force.

5

Mould Parting Line

A blow mould is necessarily composed of two halves, which move towards each other to close over the parison and go apart after completion of blowing and cooling operations. The cavity is housed in both mould blocks. The face, where the two blocks meet, is called the parting face and the edge of the mould cavity is termed as the mould parting line. It is visible on the outer surface of the blown article as a fine line.

Criteria for choosing the parting line are:

- Axis of symmetry.
- Ejection of the blow-moulded article without hindrance and distortion.
- Uniform stretching of the parison.
- Not overstepping the permissible blowing ratio.
- Ease of placing and securing inserts.

In case of symmetrical articles, the parting line divides the product in two equal halves, either identically or as mirror image of each other. The line runs through the axis of symmetry of neck or mouth, body, handles, eyelets etc. Containers for packaging of consumer goods are usually symmetrical about at least one axis with their opening situated in the centre. The course of parting line in case of articles like round bottles, jerry cans,

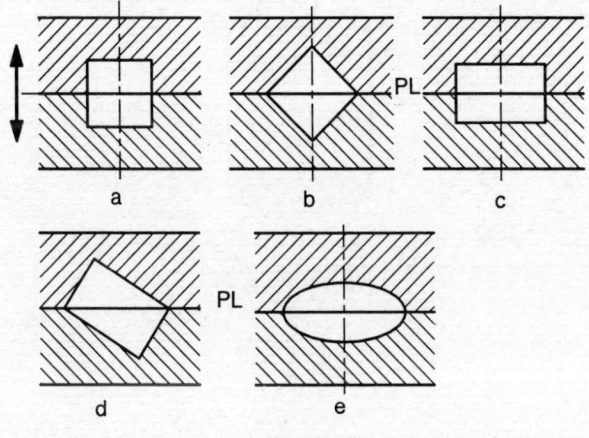

Fig. 5.1.

and bottles with handles or eyelets is quite obvious. Square or rectangular containers can be parted either in the middle of two oppositely situated side walls (shorter ones for rectangular products) or diagonally, with the parting line dividing the corners (Fig. 5.1).

Both methods have their advantages and disadvantages. In the former alternative, the matching of the mould halves is easy and the stretching ratio of the parison is favourable, though ejection of the article from the mould is slightly difficult. Conditions are just reverse with the parting line running diagonally.

Elliptical containers are parted along the major axis of the ellipse. It facilitates ejection and permits use of large parisons to obtain adequate wall thickness even at the point of maximum stretching.

Technical articles with undercuts, two or more openings not situated on the same line and other unsymmetrical features, make the choice of a zigzag parting line unavoidable (Fig. 5.2). The parting line must enable removal of the blown article without distortion or damage. Should it result in unfavourable stretching, movable inserts may have to be incorporated in the mould

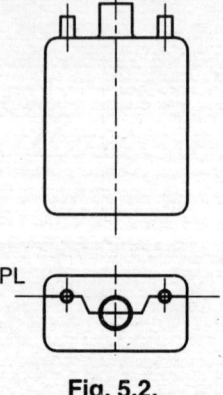

Fig. 5.2.

to combine acceptably uniform stretching with trouble free ejection. Articles with integral hinge, such as boxes with lids, may have to be parted in such a way that the parting line passes through the middle of the article but is raised up in the section where the hinge is situated. As the Fig. 5.3 shows, a straight parting line would have divided the article unsymmetrically, causing uneven stretching and unequal wall thickness. On the other hand, a nearly symmetrical article may have to be parted unsymmetrically to provide adequate wall thickness to an oddly situated feature (Fig. 5.4).

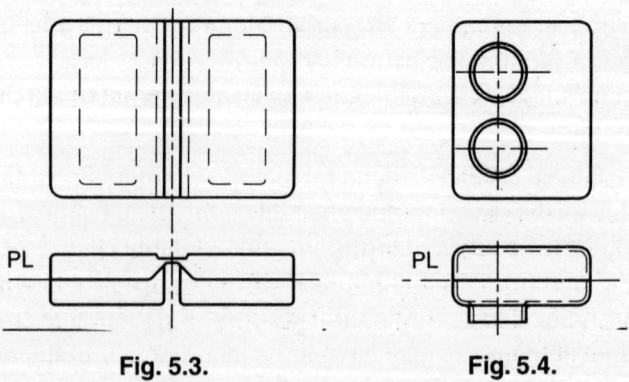

Fig. 5.3.   **Fig. 5.4.**

It will be observed that the mating faces of the cavity blocks in all cases discussed so far will be either flat in one plane or flat but stepped. Some technical articles with bends and curves make the choice of odd parting lines unavoidable. Blow moulded toys represent extreme cases. Here the ease of ejection becomes the prime factor dictating the course of the parting line. As obvious, the mating faces of the cavity blocks will no more be plain. Fig. 5.5 demonstrates parting line of an unsymmetrical article.

Positioning

As a rule, the articles are positioned in the mould in such a way that they can be formed with parisons of smallest possible diameter, giving rise to as little flash as possible. There is generally a single optimum position for symmetrical or nearly sym-

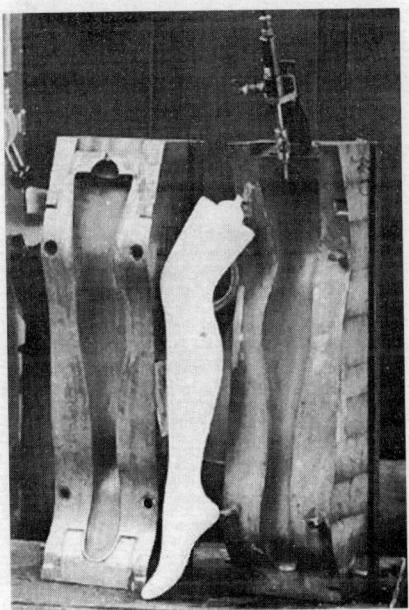

Fig. 5.5.

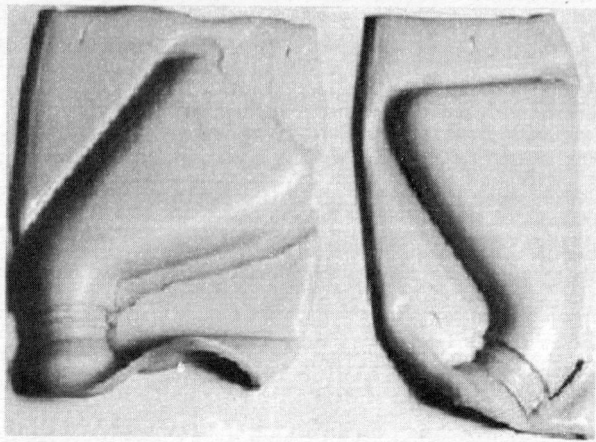

Fig. 5.6.

metrical articles. Their axis coincides with that of the mould. The unsymmetrical objects, however, may be positioned in more than one way. The criteria for selection of the most suitable position is the size of the flash as well as the least possible weld line. Fig. 5.6 compares two options. The position on the right is obviously the better solution.

CHAPTER

6

Mould Cooling

A blow mould receives a hot plastics parison, inflates it and cools it to a rigid state. Consequently, a certain amount of heat has to be dissipated with each cycle. It is achieved by circulating cooling water in a labyrinth incorporated in the mould for this purpose. Design of the cooling system in a blow mould deserves, therefore, utmost attention, not only because the cooling period may constitute as much as 90% of the blow moulding cycle, but also because of the far reaching influence of the cooling on many aspects of the quality such as surface finish, dimensional stability, freedom from internal stresses etc. Only one surface of the inflated parison is in touch with the cooled mould. Consequently, all heat in the parison has to be dissipated from one side only. It depends solely on the efficacy of the cooling network, how soon the heat can be extracted and the blow moulded object brought down to a temperature where it can be ejected without the risk of distortion of the shape. The output rate and hence the efficiency of the process are closely linked with the cooling efficiency of the mould. While laying out the cooling design of the mould, the emphasis should not only be on the rate of heat transfer but also on the uniformity of cooling as uneven cooling leads to uneven shrinkage which in turn results in generation of internal stresses and distortion.

It may be assumed that a parison has practically the same temperature at all points. As the two mould halves close over it, a certain part of it is necessarily pinched off around the neck, the base and in certain cases on the sides too, between the mould halves. This part of the parison, referred to as "the pinch off" or "the flash", does not undergo any expansion and constitutes a zone of heat concentration.

Depending upon the configuration of the article, the parison stretches to different degrees at different points. There is very little or no expansion in the neck area. The welding seam at the base also remains very thick whereas corners, belly and sections with large dimensions necessitate maximum stretching of the parison. The areas of less expansion represent zones of heat concentration. An efficient layout of the cooling network takes into account all these factors and provides for faster dissipation of heat from the critical areas so that the article as well as the flash cool simultaneously.

As mentioned in the outset, blow moulds are invariably cooled by circulating water through channels in the mould, provided specifically for this purpose. The choice of their shape, size and the layout decides efficiency of the cooling. In case of medium and large sized moulds, it is customary to provide a number of independent cooling circuits to enhance the efficiency. Areas of heat concentration, especially the neck and the base are cooled with individual circuits, except in very small moulds where the space may not permit placement of more inlets and outlets. Another reason for cooling neck and bottom inserts more intensively is the inferior heat conductivity of the alloyed steel used for fabrication of these mould parts which are subject to higher wear and tear.

There are several ways in which blow moulds can be cooled, the choice of a suitable cooling system being dependent on the design and size of the product and material of the mould. The various systems employed for cooling are:

1. Cooling Chambers

The back of the mould cavity is milled in the form of interconnected chambers to provide a circuit for water. As far as possible, the shape and the depth of the product is followed to maintain a uniform distance between the water chambers and the mould cavity. To this end, the chambers are milled in progressive steps. It is preferable to have a large number of small chambers than a few big ones to improve the cooling effect and to provide better support to the mould cavity by means of numerous partition walls, which may also be conveniently employed to house the vent holes. The cooling system is sealed by means of a gasket, placed between the mould cavity block and the back plate. The inlet and outlet holes can be housed in the back plate (Fig. 6.1).

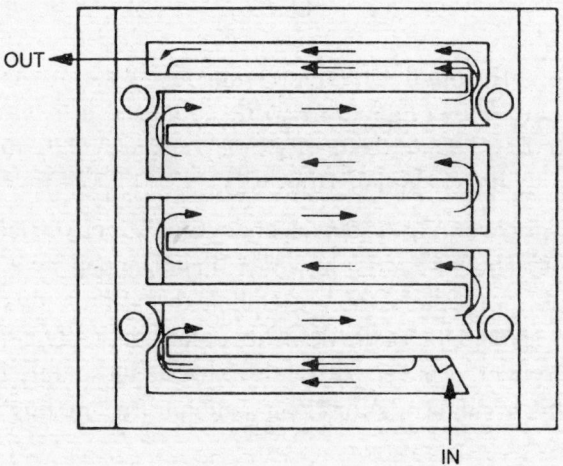

Fig. 6.1.

2. Spray Cooling

Usually employed in very big moulds for large containers, drums and tanks, the system consists of a number of pipes with a large number of small holes through which water is sprinkled against the back of the mould cavity block. A number of smaller pipes,

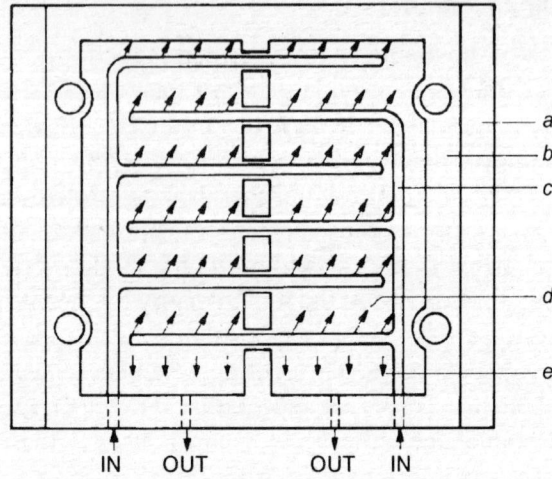

Fig. 6.2.

branching off from the main inlet pipe take water to the remotest corners. The complete network is housed in cavities machined or cast behind the cavity block and sealed by means of gaskets and the back plate (Fig. 6.2).

The spray cooling system is, undoubtedly, very efficient and adaptable. The system is, however, handicapped by a serious drawback. Some holes may get choked and thus give rise to local hot spots. Unless the defect becomes optically noticeable on the product, it is very difficult to locate its source. In other words, the system calls for clean, calcium free cooling water.

3. Drilled Cooling Lines

A number of vertical holes, placed around the cavity and connected alternately by means of cross holes or milled channels, form a cooling network, which represents a cheap but fairly effective cooling system for small and medium sized blow moulds. The centre distance of the holes from the cavity surface is chosen as 1,5 times their diameter. A gap equalling one hole diameter is left between two adjacent holes. Holes of sizes less than 8 mm and more than 15 mm are seldom used. Efficacy of the system can be enhanced by introducing spiral shaped copper strips in the holes to create turbulence (Fig. 6.3. See also Fig. 3.1, chapter III).

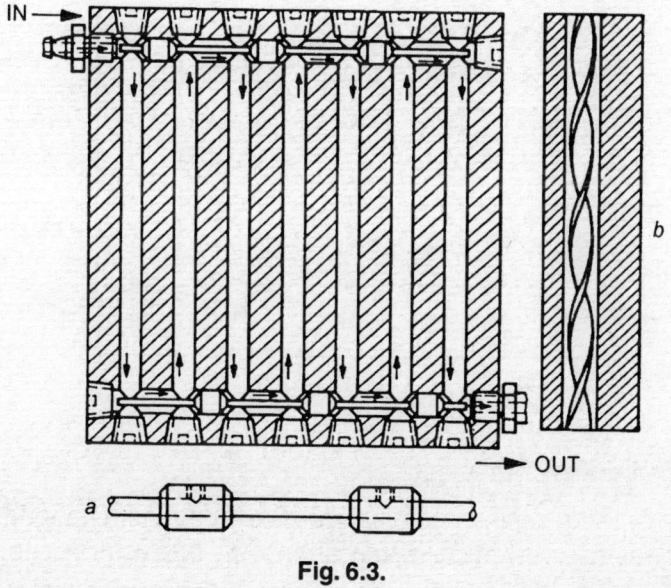

Fig. 6.3.

The system is less effective if the article to be blow moulded has an irregular shape as the straight cooling holes will run at varying distance from different sections of the article.

4. Zigzag Cooling Lines

The system comprises a network of zigzag cooling holes at the back of the mould cavity, following the contour of the product. To create the network, the back of the mould block is hollowed, leaving sufficient material behind the cavity. A wax core, following the shape of the article, is laid out in the hollowed chamber, which is now filled with a metal powder impregnated thermosetting resin. Wax is melted and blown out after the resin is cured.

This design (Fig. 6.4) is eminently suitable for cooling moulds for articles having irregular shapes.

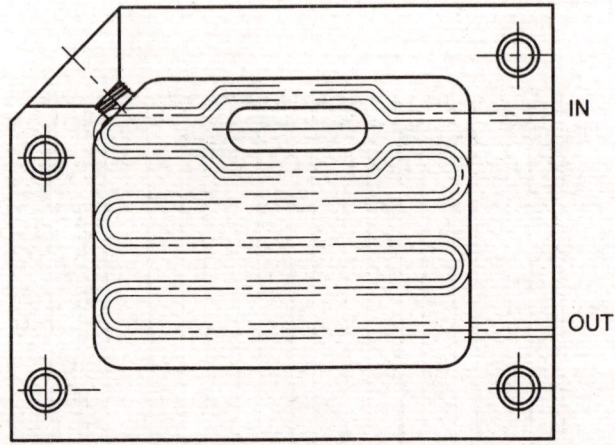

IN

OUT

Fig. 6.4.

5. Embedded Tubes

A network of copper tubes formed to cover the configuration of the article, is cast in the mould at the same time as the cavity is formed. The effect of the cooling may be compared to that of the zigzag lines. The layout of cooling pipes can be made to follow the form of the product. It must be remembered that drilled or cored holes in cast moulds may lead to leakage of water through interconnected blow holes. Embedded pipes eliminate this danger (Fig. 6.5).

Copper tubes inserted in drilled or cast holes form an ineffective way of cooling. The overall heat transfer from the mould to cooling water flowing through the copper pipes is inefficient because of the air gap between the pipe and the surface of the drilled hole.

6. Cooling with the Blowing Mandrel

The methods described so far cool only the outer surface of the

Fig. 6.5.

blown object. The blowing air introduced through the mandrel inside the parison can also serve as a coolant if it is continuously renewed. The mandrel can be designed to let out a controlled amount of air through a by-pass, maintaining the required air pressure at the same time (Fig. 6.6).

Cooling with flushing air as described above proves less effective with large products where the mandrel may be far away from the other end of the article. The flushing air may form a short circuit, not effecting the remote parts. With most technical articles like pipes for exhaust air of automobile engines, water pipes in dish washing machines and other such products, the ends are moulded closed and cut off in post operations. With such articles, internal cooling can be improved by providing an exhaust at the remote end through a piercing needle.

The neck of the article, which constitutes a zone of heat concentration can also be cooled from inside with the help of the blowing mandrel if the diameter is not very small. A cooling circuit can be built in the mandrel, through which water can be circulated. (See also Chapter X, Design of Blowing Mandrels)

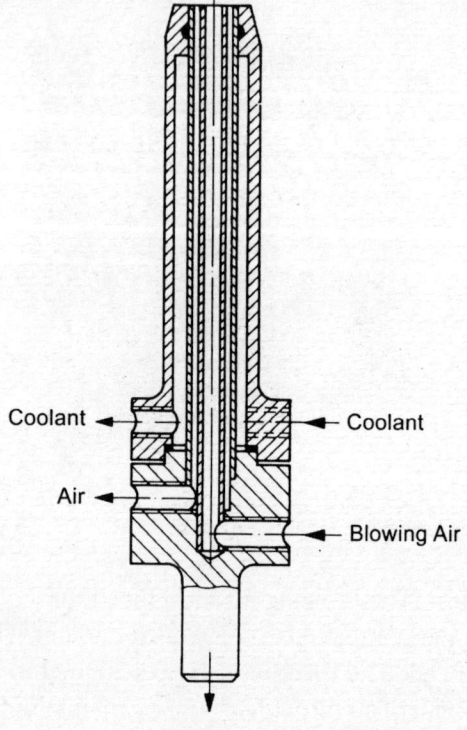

Fig. 6.6.

For thick walled articles with sufficiently large openings, the blowing mandrel can be provided with arrangement to introduce liquefied Carbon Dioxide after inflation for cooling the article from inside. The method is very effective and helps cut down the cycle considerably. However, it calls for additional equipment for regulation of the gas at sub-zero temperatures. The pipes and the mandrel have to be protected against icing. Nitrogen, too, can be employed for faster cooling of the thick walled articles but the high cost of these gases and of the additional equipment and arrangements proves economical only in exceptional cases. Experiments with blowing air cooled at different temperatures, on the other hand, have yielded more positive results and have led to the conclusion that the blowing air at minus 20°C and at a pressure of 10 bars can transport away as much as 30% of the heat from the article. A more practical compromise is the use of air at 20°C. It can dissipate about 10% of the heat from the article being blow moulded.

The compressed air for blowing must be free from oil, moisture and other impurities.

A cooled blowing mandrel with provision for air circulation can dissipate 25-40% of the parison heat.

7. Cooling with Air Jets

Whereas the pinch off caught between the mould halves gets cooled by the cavity blocks and mould inserts for top and bottom section of the mould, the surplus part of the parison projecting out of the mould may remain hot enough to stick to the blown article upon ejection. Pipes, with a number of tiny holes, fixed parallel to the mould parting line on both mould halves can be employed to cool the flash by means of air jets (Fig. 6.7).

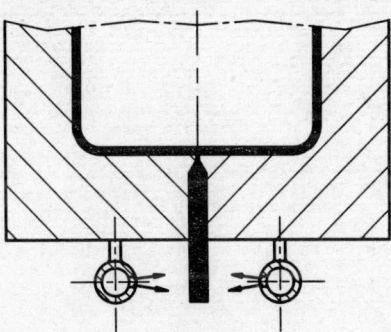

Fig. 6.7.

In conclusion, it may be noted that chilled water also expedites cooling and helps reduce the total moulding cycle. The rate of heat removal, however, is not directly proportional to the temperature of the cooling water because of the poor thermal conductivity of plastics. The thermal conductivity of different plastics is also different. The heat dissipation from thicker sections slows down after formation of a frozen layer.

Too cold water may lead to condensation of moisture on the mould which, in turn, can cause corrosion in the mould and orange peel and watermarks on the product.

The ideal mould temperature is not same for all thermoplastics. For every polymer, the raw material manufacturers recommend a mould temperature range, which would ensure optimum quality. In fact, the mould may have to be heated in case of engineering thermoplastics, which are processed at high temperatures. However, the blown product is always being cooled in spite of the seemingly high temperature of the mould.

The ideal mould temperature for different materials are listed in Table 6.1

Table 6.1

Material	Mould temperature °C
L.D.Polyethylene	20 25
H.D.Polyethylene	8-30
Polypropylene	6-25
Polyacetal	90-100
G.P.Polystyrene	20-30
H.I. Polystyrene	20-30
Acrylonitrile Styrene	20-30
ABS	20-30
Cellulose Acetate	50-60
PVC Rigid	20-50
Polycarbonate	80-100
Polyamides	70-90
Polyethylene Terephthalate	25-30 (Injection blow moulding)
Glass polymer	10-38
Polysulphone	150

CHAPTER

7

The Pinch Off

Pinch off is the section of the mould where the parison is gripped between the mould halves. It performs three main functions:

- Welding of the open ends of the parison.
- Separation of the flash or the superfluous material from the article.
- Cooling of the flash.

For blow moulding symmetrical articles with their opening on the axis of symmetry, the parison is invariably smaller than the breadth/ diameter of the moulding. Even in the folded state, it does not go beyond the periphery of the article. Consequently, welding of the parison and cutting off of the flash is confined around the neck and the shoulders on one side and below the base on the other side. Fig. 7.1 illustrates these details.

Unsymmetrical products, however, may call for a parison, which may be required to cover not only the neck and the bottom but also the sides, either fully or partially. Jerry cans, containers with neck on one side or at an angle, bottles with side handles, articles like balls, capsules, vials etc. which may be formed out of a single parison in batches, are a few examples of such cases. Here the pinch off is required all along those sections where the parison extends beyond the cavity. Technical

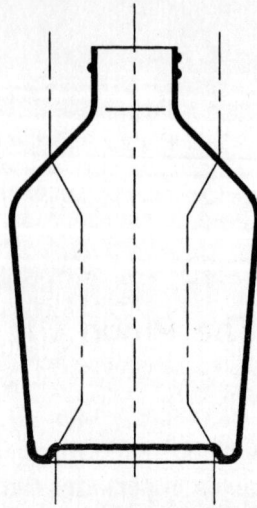

Fig. 7.1.

articles like air ducts and fluid pipes in automobiles, having 3D configurations, also fall in this category (Fig. 7.2). The folded parison, in such cases, is required to cover the whole breadth of the component and the flash may outweigh the product by 100% or more.

(Some special blow moulding techniques and machines have been developed to circumvent this drawback, especially in case of technical, pipe-like articles bent in three dimensions. Two of them, viz. the Robot-aided Parison Placement and Suction Blow moulding, have been dealt with in chapter XVIII for special blow moulding processes).

The simplest design of the pinch off consists of a narrow edge parallel to the contour of the particular article section (e.g. the bottom), sloping outwards after certain flat breadth (Fig. 7.3). Another configuration includes a second flat section for compression after the slope (Fig. 7.4). It squeezes the overhanging parison and pushes it partly inwards to thicken the welding seam. It also cools the flash more effectively. Fig. 7.5 depicts

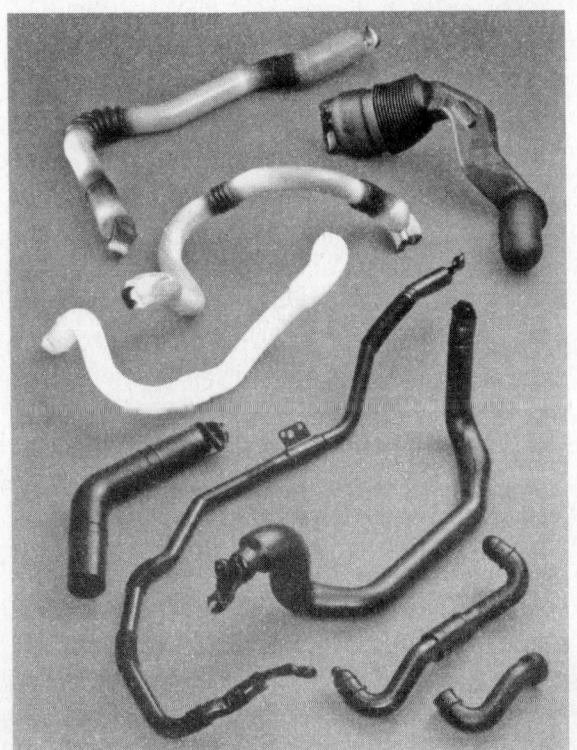

Fig. 7.2.

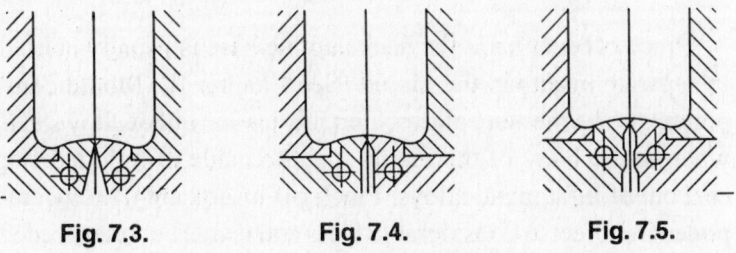

Fig. 7.3.   **Fig. 7.4.**   **Fig. 7.5.**

yet another design used in difficult cases. The constriction or
the dam helps to create a strong welding seam by pushing up
the melt into the cavity along the pinch off line. Fig. 7.6 illustrates the difference between a weak and a strong welding seam.

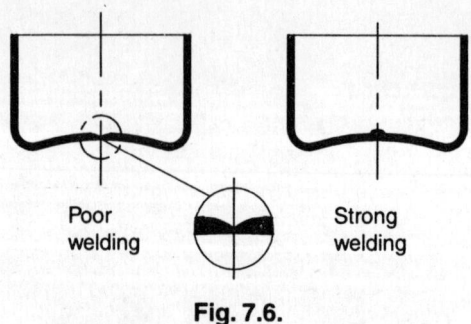

Fig. 7.6.

Another version of the pinch off, employed mostly for closed features like integral, hollow handles, contains yet another flat section (Fig. 7.7). The first flat pushes the melt up to strengthen the welding seam and the second flat cools the flash.

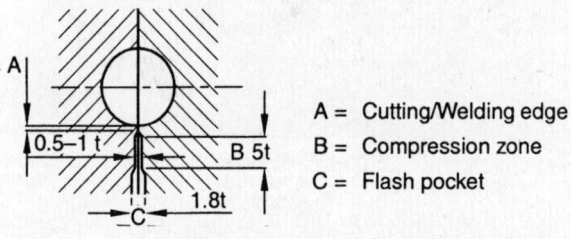

A = Cutting/Welding edge
B = Compression zone
C = Flash pocket

Fig. 7.7.

Pinch off section at the neck and the base is usually housed in separate inserts in the mould (See Chapter III, Mould components). The measure allows their fabrication out of alloy steels whereas the body of the mould may be made of mild steel or cast out of light metal alloys. Pinch off inserts are mould components subject to considerable wear and tear. The force needed to effect a neat separation of flash may range between 1000 N per cm. of the pinched length for L.D.Polyethylene to 2000 N or more for engineering plastics. Due to this reason alone, they must be interchangeable. Another advantage of inserts, however, lies in the ease of their independent and intensive cooling, which is essential to dissipate the heat from these areas of heat concentration. Yet another beneficial effect of inserts is their contribution to venting (See also chapter VIII, venting).

The breadth of the welding edge, the degree of slope and the width of the compression section decide the strength of the welding seam, the quality of flash separation and the rate of cooling of the flash. Whereas the width of the compression zone is always chosen as 9/10th.of the parison thickness, the breadth of the welding edge and the degree of slope vary with the size and material of the moulding. The thickness of the film left between the blown article and the flash should not exceed 0.1 mm. An ideal film is translucent and can be easily frac- tured.

In order to expedite cooling of the flash, especially in case of thick parisons which may sometimes take longer to cool than the article, the pinch off area is provided with furrows (Fig. 7.8). It enlarges the surface area of the flash held between the cooled pinch off inserts. It may also be used for automatic sepa- ration of the flash in the mould itself. Fig. 7.9 illustrates the practical use of this device for a large article.

Table 7.1 contains the data essential for design of the pinch off. It may, however, be remarked that the welding edge will have to be considerably broader if the mould cannot be slowed down in the last phase of closing.

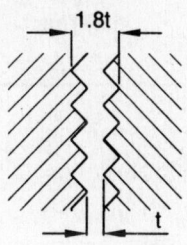

Fig. 7.8.

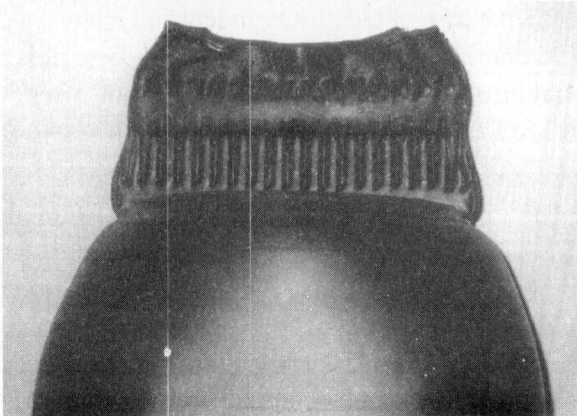

Fig. 7.9.

Table 7.1

Material	Breadth of the the welding edge	Included angle (in degrees)
Polyethylene, low density	0.2-4 mm	15-30
Polyethylene, high density	0.2-4 mm	15-45
Polypropylene	0.3-4 mm	15-45
Polyvinyl Chloride	0.5 mm	60
Polystyrene	0.3-1 mm	25-30
ABS	0.3-1 mm	30
Polyacetal	0.5 mm	30
Polyamides	0.3-4 mm	30-60
Polycarbonate	0.2-2.5 mm	30-40
Cellulose Acetate	0.5-1 mm	45
Glass Polymer	0.2-0.3 mm	60-90

Venting

An empty mould contains air, which must be expelled completely so that the expanding parison can assume the shape of the cavity cut in the mould in all details without any resistance. The process of expulsion of air is performed by the parison itself which, by virtue of its higher internal pressure, drives the air out as it is inflated, provided there are no air traps and the means of escape are present at the points towards which the enclosed air will be driven. These escape routes are referred to as the air vents. Their importance can be comprehended readily when one considers the fact that the mould represents the only major means of dissipation of heat that is brought in by the hot parison with each moulding cycle. Should the trapped air, which is a bad conductor of heat, prevent the parison to sit snugly on the cavity walls at all points, the resulting uneven cooling will give rise to a number of defects such as uneven shrinkage and hence stresses and distortion, poor reproduction of the mould details and dimensions, uneven surface finish, pit marks, orange peel besides longer cooling period. The result can be summed up in one word: "bad moulding". Design of venting merits, therefore, due consideration.

Before going into the design details of venting, it will be advisable to investigate where the air is most likely to get trapped. As the parison expands gradually, it pushes out air through the mould parting line where the two mould halves meet. This path of escape, however, gets sealed very soon as it is situated close to the folded parison. The parison touches the parting line in the beginning of inflation and blocks the path of air. Wherever the expanding parison touches walls meeting each other such as the shoulder section, the rim and the grooves and engravings etc., it forms an air trap. All corners, undercuts, depressions, engraved scripts and logos, threads and beads constitute such traps. In other words, these are the critical areas where provision of air vents is essential. Various methods adopted to this end are:

a) Venting through the parting line

Parting line venting is the easiest way of providing escape paths to the air in blow moulds. It proves adequately effective in case of small moulds with simple cavities. A number of slots, 0.05-0.1 mm deep and 10 mm broad are machined on the parting face of one of the mould halves. The slots are deep enough for the enclosed air to escape but too shallow to leave any unsightly impression on the surface of the product. To enhance their effectiveness, the grooves may be deepened after a distance of 1.5-2 mm from the edge of the mould cavity (Fig. 8.1). With parting line venting, approximately 50% of the mould parting face is vented.

Venting is also provided by the base and shoulder inserts through their joints if the parting line for the insert is judiciously selected. Firstly metal against metal never forms an air tight joint and secondly, additional venting slots can be provided on the mating surface of the inserts. The corners, which invariably form air traps, can thus be effectively vented. However, additional means of venting may be found necessary for larger moulds.

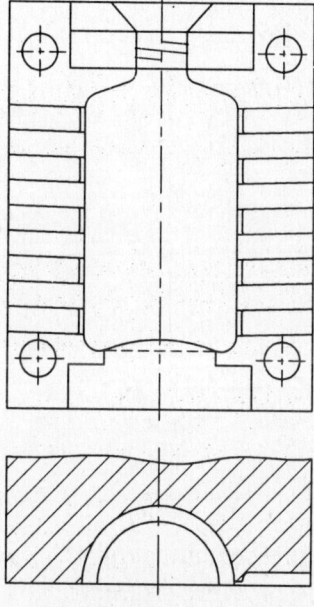

Fig. 8.1.

b) Venting holes

For moulds having critical design details, threads, grooves, depressions and such other traps for air, vent holes at critical junctures provide the most effective means of venting. Relatively large holes are drilled from the back of the cavity to a distance of 0.5-1.5 mm of the mould cavity surface. From this point, very small holes, 0.2-0.3 mm. in diameter, open up in the cavity (Fig. 8.2). The number of such vent holes depends upon the size of the article, the number of air pockets and the speed of production. Their size, however, should not exceed the given limit for obvious reasons.

The threads around the neck of blow moulded containers are invariably housed in inserts foreseen with cooling grooves, which come in the way of venting holes. Here, the small venting holes are diverted before reaching the cooling grooves.

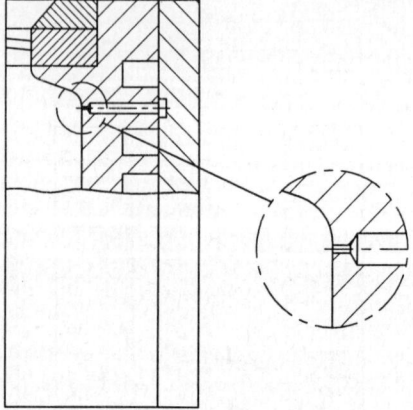

Fig. 8.2.

c) Venting Plugs

A hexagonal plug, pressed into a round hole drilled at the spot of entrapment of air, provides six vent apertures with a camber height varying between 0.1 and 0.2 mm (Fig. 8.3). The method is particularly adaptable for venting of the areas where drilling tiny holes may not be feasible. The number of holes and plugs will depend upon the size of the mould and the surface finish of the blown article. A larger through hole under the plug lets the vented air out of the mould.

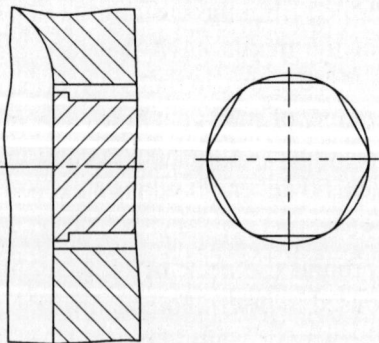

Fig. 8.3.

Flat, shallow mould cavities with a depth upto 25 mm. can be adequately vented with one 3 mm.diameter hole (with a hexagonal plug in it) per 25-30 square centimetres of the mould area. For larger moulds, bigger plugs may be employed if their impression at the particular spot is not found disturbing. The camber height, however, must remain within the specified limits.

d) Standard Venting Inserts

A variety of prefabricated plugs, mostly in brass or light metal alloys, with fine venting slots or tiny holes, available readymade, offer a broad choice for venting of blow moulds of different sizes (Fig.8.4). They can be fitted in a stepped hole like the venting plugs before sandblasting of the cavity, thus leaving no unsightly impression on the surface of the product. The hole at the back should be sufficiently big to provide a free path to the air passing through the slots or small holes of the plug.

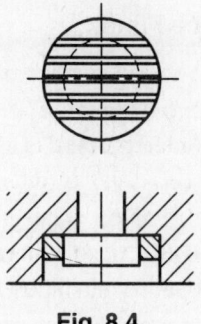

Fig. 8.4.

e) Porous Metal Inserts

Sintered metal inserts can be employed for venting of moulds with deep engravings and intricate designs. The porous metal provides an effective path of escape to the air that is bound to get trapped in engravings. A number of holes in the mould behind the inserts form the final route of air escape (Fig. 8.5). Porous metal inserts with proper grain size provide effective as

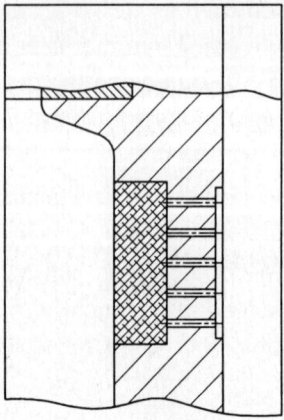

Fig. 8.5.

well as inexpensive venting device without influencing the surface finish of the product adversely.

f) Mould Cavity Surface Finish.

Minute depressions of a rough surface of the mould cavity form
a path, through which the trapped air can bleed out to the vent
holes, even when the inflated parison is touching the cavity walls.
This is one of the reasons why blow moulds for most of the
thermoplastics are sand blasted. The grit size of sand for normal surface finish is 60-80 mesh whereas finer sand particles
are used for blasting of cavity surfaces for articles, which have
to be decorated. The air pressure for sand-blasting is confined
to 1-1.5 bar.

Undercuts

Undercuts in a blow moulded article may be defined as features which, by virtue of their shape, size or position, oppose release of the article from the blow mould.

Hollow articles, even those moulded out of rigid thermoplastics, possess a certain amount of flexibility because of their hollowness. Undercuts, to a certain extent, can therefore be surmounted without special aids or additional devices in the mould. As flexibility is an attribute dependent on many factors like rigidity of the material, size and shape of the article and its wall thickness, it is difficult to lay down hard and fast rules regarding undercuts, which will apply universally. Some useful data will be found in chapter XVI (Design of Blow Moulded Containers). It cannot, however, be denied that some undercuts make incorporation of special releasing and ejection devices in the mould unavoidable.

In the outset, three factors have been named which are responsible for undercuts. These are shape, size and position. Whereas some undercuts may be attributed to only one of the three factors, it is quite likely that a combination of two or all three of them may be found present in most cases.

Threads and grooves inside the neck of a container will make the blown article stick onto the blowing mandrel. Ejection by stripping is feasible only when their depth does not exceed a certain limit in relation to the size and the thickness of the opening and to the flexibility of the material. For normal threads, these limits will be found impractical. The most common solution adopted to form internal threads and have a safe ejection as well is to employ a blowing mandrel with a detachable tip bearing the desired threads. The tip is ejected along with the product and unscrewed out of it outside the mould. In the meanwhile, a second identical tip is substituted for the next cycle. A similar procedure is followed if the threads are situated not inside the neck where blowing is taking place, but in another depression situated on the parting line or on a side-wall. Dummy mandrels or loose cores are placed in position and ejected along with the blown product. External threads too form an undercut unless situated on the parting line. Should their position make it impossible to include them in the path of the parting line, which can run zigzag for this purpose, loose inserts are employed to form these threads on the article. It goes without saying that the inserts will be ejected after every cycle with the blow moulded product, removed from it and replaced in the mould.

The additional operation of placement of cores and inserts in the mould and subsequent unscrewing outside can be eliminated in two ways. Cores can be incorporated in the mould and unscrewed by rotation with the help of hydraulic cylinders or motors before the mould is opened. Although the process becomes automatic and continuous, the moulds become complicated, more expensive in fabrication and in maintenance. Only very large series of production justify this step.

Metal or plastics inserts with internal threads are increasingly being used instead of loose cores to eliminate the post operation of unscrewing. These can be secured on the blowing mandrel, on dummy mandrels or on cores incorporated in the mould for this purpose (See also chapter XVI).

Handles and grips are an essential feature of large sized containers. Jerry cans are invariably provided with integral handles. For the sake of stackability, either the handle and the neck are placed in a depression to make the top face flat for secure placement of another jerry can on it or the base is domed inwards to accommodate the handle and the neck of the container below. Both measures give rise to undercuts, which make release of the blown container from the mould without the aid of special devices impossible. Mould components corresponding to the depressions on top or base of the article form solid obstructions, over which the product cannot jump without permanent deformation or damage as the mould opens. The only way out is to remove the obstruction before or during the opening stroke.

To achieve this end, either the base or the top part of the mould is designed as an independent unit with its own cooling circuits, joined to the main body of the mould by means of guiding members which allow it a free sideways movement as the mould opens (Fig. 9.1). It results in lifting up of the article gradually in accordance with the space created on the other end. The article gets released without damage or distortion. The other alternative for clearing the obstruction is to pull back the inserts having the undercut before opening the mould (Fig. 9.2).

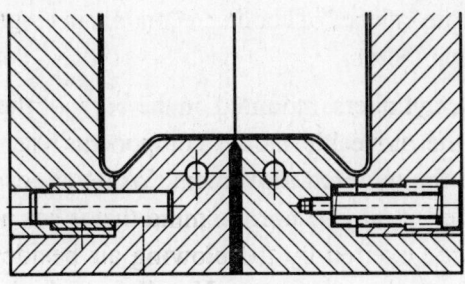

Fig. 9.1.

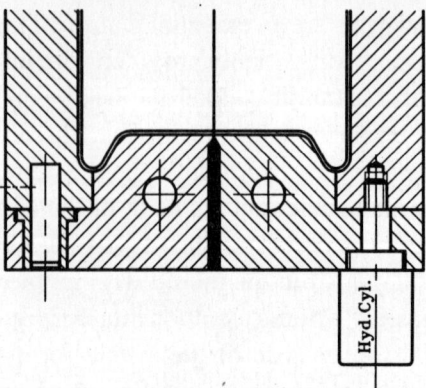

Fig. 9.2.

In moulds of larger sizes, it is advisable to employ four guide pins and bushes for each side. The movement becomes easier and smoother if ball cage bearings are used instead of conventional bushes.

It may be remarked that it is not necessary to make the particular side having the undercut moveable, unless the body of the article possesses features which prevent its up and down movement. As the mould opens and base/top make a vertical movement creating room for the blown article to rise over the obstruction, the article can adjust itself. The criteria for the choice as to which components should be made moveable, is provided by the position of the article in the mould with respect to the cross head, convenience of cooling, economy in length of parison and other such factors.

Hydraulic cylinders, mounted on the body of the mould and attached to the moveable mould components viz. base or the top, provide the required movement. While choosing the size of the cylinder, it may be born in mind that it has also to withstand the force exerted by the blowing air besides having to move the specified mould parts. Usually, one hydraulic cylinder placed in the middle of each moving component suffices. However, if it is not possible to accommodate the cylinder in the middle, it is advisable to use two smaller ones placed symmetrically.

Pneumatic cylinders of conventional construction cannot bring up sufficient force to counterbalance the total force exerted by the blowing air from within. However, those with self-locking arrangement may be considered for the job if their cost does not outweigh that of the hydraulic cylinders.

Mould Shrinkage

Thermoplastics expand in volume on melting and therefore have a lower specific gravity in this state as compared to that in the solid form at room temperature. The subsequent contraction in volume upon cooling leads to a decrease in dimensions, which is termed as mould shrinkage. The change in volume and hence in dimensions is much greater in case of crystalline thermoplastics like the Polyolefins and the Polyamides than with the amorphous ones like Polystyrene or Polyvinyl Chloride.

Mould shrinkage may be defined as the ratio of the dimensional difference between the moulding and the mould, measured 24 hours after moulding, at room temperature. Dimensions of the mould for a proposed article have to be chosen larger by the amount of shrinkage, which is different for different plastics. In mathematical form, the mould shrinkage may be expressed as below:

$$\text{Mould shrinkage (in \%)} = \frac{\text{Mould dimension-the article dimension}}{\text{Article dimension}} \times 100$$

Experience has shown that certain processing parameters viz. mould temperature; cooling period and the stretching ratio exercise a marked influence on the shrinkage of a blow-moulded article. An increase in the mould temperature or in the stretching ratio leads to a higher shrinkage, whereas a longer cooling period results in decrease of the extent of shrinkage. Change in temperature of the melt and in pressure of the blowing air, on the other hand, do not cause any appreciable variation.

Plastics are invariably poor conductors of heat. This property, coupled with the fact that heat is dissipated practically from one side of the moulding only, leads to faster cooling of thinner sections of the moulding. Unless the cooling period is very long, the thicker sections of the ejected moulding will have higher temperature than the thinner ones. While cooling without confinement and pressure, the thicker sections undergo higher post shrinkage.

Crystalline thermoplastics display differential shrinkage behaviour. They shrink more in the radial direction than along the axis. In other words, the diameter of a round bottle out of Polyethylene, a semi crystalline material, will shrink more than its height. Due to stretching, the molecular structure gets oriented in direction of deformation. This difference is also termed as the orientation shrinkage. Stored mouldings undergo further shrinkage because of the rearrangement of the macromolecules. The post shrinkage is far more remarkable with semi crystalline thermoplastics than with the amorphous ones.

It all goes to show that shrinkage of plastics cannot be compared with the coefficient of linear expansion of metals. Distinction must be made between the radial and axial dimensions, thick and thin sections and areas of more or less expansion while calculating the mould shrinkage. This is why, the shrinkage of plastics is always stated as a range and not as a definite figure. The raw material manufacturers quote mould shrinkage values for their polymers, determined in the laboratory as per standard procedures.

Table 10.1 lists linear shrinkage of some thermoplastics commonly used for blow moulding.

Table 10.1

Material	Shrinkage %
Polyethylene, Low Density	1.5-3.0
Polyethylene, High Density	1.5-5.0
Polypropylene	1.5-2.5
Polyvinyl chloride	0.5-0.8
Polystyrene, SB	0.6-0.8
ABS, SAN	0.6-0.8
PMMA	0.5-0.8
Polyacetal	1.0-3.0
Polyamides	0.5-2.8
Polycarbonate	0.5-0.8
Cellulose Acetate	0.6-0.8
PET	0.2-0.4
Glass Polymer	0.3-0.6
Polysulphone	0.7

11

Blowing Mandrels

The primary aim of a blowing mandrel is to introduce blowing medium viz. compressed air into the plastics parison in order to inflate it to a predetermined shape represented by the mould cavity. Modern techniques have, however, assigned to it other functions also which influence its shape and construction. The rules of its design can be comprehended readily if various functions and forms of blowing mandrels have been made clear in the outset.

Whereas a simple blowing mandrel has no direct relation with any dimension of the product being blow moulded, a calibrating mandrel, as mentioned before, forms the internal diameter of the neck. The "Kautex" methods of calibrating from below makes use of a vertical mandrel, which can be adjusted horizontally and can move up and down. It is inserted into the parison before closing of the mould over it. There is no resistance to its entry either from the side of the parison, which is bigger in diameter than the mandrel, or from that of the mould which is then in the open state. On closure of the mould, the parison is squeezed between the mould halves and the mandrel, superfluous material being thus forced out into flash pockets provided on both sides of the neck. The neck, held between the moulds and the mandrel, forms an effective seal against leakage of the compressed blowing air.

As the mould opens at the end of a cycle, the blow moulded article holds onto the mandrel and is stripped off from it when the mandrel makes a short stroke downwards and the article is hindered in accompanying it by a stripper plate, fixed around the mandrel below the mould.

In order to understand the secondary functions of a blowing mandrel, it will be necessary to touch, once again, upon the subject of cooling. The absence of a mould core like that in injection moulding makes the process of cooling of the article one-sided as heat can be extracted only from outside of the moulding through the cooled mould surface. The poor heat conductivity of plastics makes long cycles inevitable unless some additional means of cooling are adopted to expedite the rate of heat dissipation. Neck of a blow moulded article represents invariably a zone of heat concentration and the calibrating mandrel, if cooled itself, can perform the role of a core in an injection mould and conduct away the heat from the section it is in contact with.

Another source of internal cooling can be provided by the blowing medium viz. the compressed air itself, if it is continuously renewed. In other words, the blowing mandrel should not only introduce air into the article being blown but also provide an escape to it in such away that the required level of pressure is maintained inside the article and fresh air finds entry continuously. Cooled blowing air can further expedite heat dissipation through convection.

A "Kautex" type calibrating mandrel, as depicted in Fig. 11.1, consists essentially of a straight shaft whose outer diameter corresponds to the internal diameter of the article's neck. It controls only this dimension and has no influence on the outer diameter of the neck, which is formed by the mould or on its height, which is limited by the cutting ring in the mould. Its diameter has to match with the inner diameter of the cutting ring and with that of the supporting projections on the top insert (See chapter 3).

This type of mandrel can also be employed for blowing of products with domes or "lost head", whereby its diameter does not match with any of the product dimensions.

A "plunge neck" mandrel, on the other hand, possesses two diameters. The smaller diameter controls the inside dimension of the neck and the larger one forms its top edge including the outer diameter and the height of the neck. It will be observed that the finish of the top of the neck will differ with different types of mandrels. In case of the former mandrel, the top edge will bear a parting line, a mark left by the two cutting ring halves. The top face, therefore, cannot be expected to be absolutely flat. Formed by a plunge neck mandrel, the top face displays no such blemish. This aspect may assume importance when the top face of the neck is required, in conjunction with a plug or a cap, to make the container leakproof.

The "plunge neck" mandrel is also referred to as the calibrating mandrel with shear ring.

The various functions, which a blowing mandrel has to perform, may be summed up as:

- Inflation of the parison.
- Forming inside of the neck.
- Forming top of the neck
- Cooling the neck.
- Circulation of the blowing medium as coolant.
- Ejection of the blow moulded article.

A good design of blowing mandrel should do justice to above requisites according to their respective importance.

Design of a Calibrating Mandrel

Fig. 11.1 depicts a blowing mandrel in all its details. The inflating air is introduced through the central pipe whereas the return passage for the flushing air is provided by the annular gap between the inner and the outer pipe. The amount of the flushing air and consequently the blowing pressure can be regulated with the help of a stud which closes the small escape hole to various degrees.

The annular gap formed between the outer air pipe and the bore in the body of the mandrel is divided into two chambers by means of thin strips or round wires, brazed onto the outer air pipe. This measure ensures that the incoming fresh water can reach the exit only after making the full circuit and cooling the entire length of the mandrel shaft.

The tip of the mandrel is tapered for easy entry into the parison. The included angle of the tip may vary from 10 to 40 degrees and its length from 10 to 20 mm., depending upon the diameter of the mandrel. Sharp edges and corners should be rounded off. Should the mandrel be employed for calibrating the neck, its outer diameter $d1$ should correspond to the inside diameter of the neck (with shrinkage allowance added onto it).

The total length, the clamping flange diameter $d2$ and the centring spigot diameter $d3$ are dictated by the design of the blow moulding machine as also the method of securing the mandrel to the blowing unit or the device of holding the stripper ring to its carrier. The bore in the stripper ring may be larger than the mandrel diameter by 0.5-1 mm. The bore is generally flared up downwards (total angle 30-60 degrees) for positive entry of the mandrel.

With due consideration to the demands put on a blowing mandrel, mild steel can be used for its body though a case hardening steel would be preferable for a longer service life. Stainless steel is recommended for processing of PVC. Copper or stainless steel pipes, about 1mm thick, may be used for building the air passages and the cooling chambers. The stripper ring should be fabricated out of a material softer than that of the mandrel. Brass, bronze or zinc alloys are all equally suitable.

The plunge neck mandrel (Fig. 11.2) possesses basically all features of a normal blowing mandrel with the addition of a

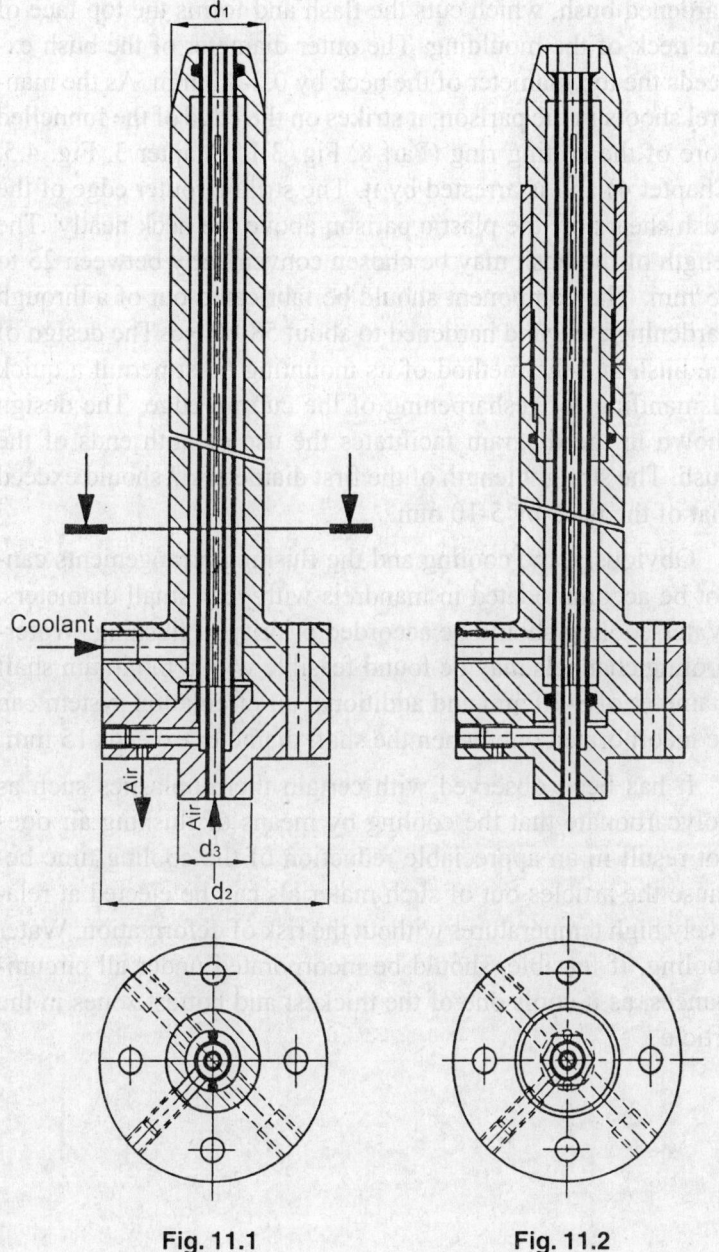

Fig. 11.1   Fig. 11.2

hardened bush, which cuts the flash and forms the top face of the neck of the moulding. The outer diameter of the bush exceeds the top diameter of the neck by 0,2-0,3 mm. As the mandrel shoots in the parison, it strikes on the edge of the funnelled bore of the cutting ring (Part 8, Fig. 3.1, Chapter 3, Fig. 4.5, Chapter 4) and is arrested by it. The striking outer edge of the bush shears off the plastic parison above the neck neatly. The length of the bush may be chosen conveniently between 25 to 35 mm. The component should be fabricated out of a through hardening steel and hardened to about 58-60 Rc. The design of the bush and the method of its mounting must permit a quick dismantling for resharpening of the cutting edge. The design shown in the diagram facilitates the use of both ends of the bush. The straight length of the first diameter $d1$ should exceed that of the neck by 5-10 mm.

Obviously, the cooling and the flushing arrangements cannot be accommodated in mandrels with very small diameters. Water-cooling should be accorded priority to flushing. Water-cooling channels may be found feasible with a minimum shaft diameter of 8-10 mm and additional air circulation system can be incorporated only when the shaft diameter exceeds 15 mm.

It has been observed with certain thermoplastics such as Polycarbonate that the cooling by means of flushing air does not result in an appreciable reduction of the cooling time because the articles out of such materials can be ejected at relatively high temperatures without the risk of deformation. Water cooling, if feasible, should be incorporated under all circumstances, as it cools one of the thickest and hottest zones in the article.

Die, Core and the Parison

It will be straying too far from the theme of this book if we go into the design details of die and the core for production of a parison. They are rightly regarded as extension of the cross head of the blow moulding machine and their design postulates knowledge of data beyond the scope of a mould designer. Their design is, therefore, always specified by the manufacturer of the blow moulding machine. It will, however, facilitate comprehension of various factors governing calculation of the parison size and choice of appropriate die and core if the basic features of the common types of these components shaping the parison are discussed here.

1. Die with Cylindrical, Parallel Annular Gap (Fig. 12.1)

This configuration represents the classical form of the die and core for production of a parison. As evident from the diagram, the starting diameters of core as well as of the die bore are dictated by the machine parameters. Both diameters are tapered down to the inner and outer dimensions of the parison. The degrees of taper, however, are kept different to create compression in the melt so that partings in the flow, caused by the legs of the spider holding the mandrel, fuse well. The total included

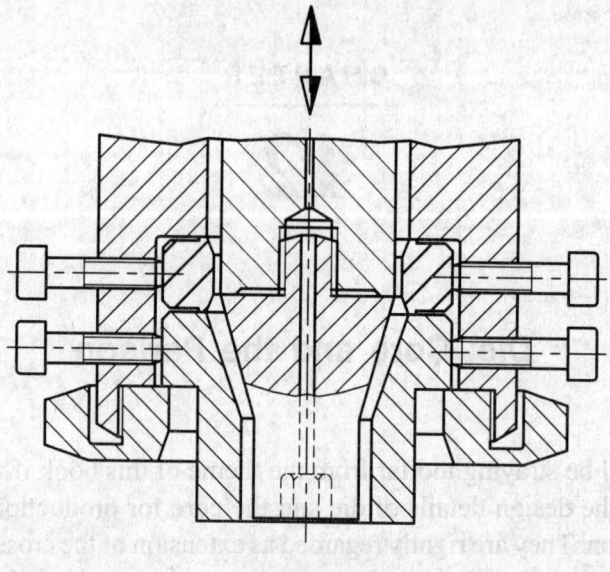

Fig. 12.1.

angle of the funnel-like bore in the die ranges between 30 to 60 degrees. The angle of the core frustum is less by 10-15 degrees.

Apart from the diameters of the die bore and the core, the other dimension which may have to be chosen by the designer, is the length of the flat section or that of the "land". It depends upon the viscosity of the material being processed as well as on the wall thickness of the parison and may vary between 5-20 times the die gap. It will be on lower side for more viscous polymers like unplasticised PVC and higher for easy flowing plastics like polyamides. Again, it is more for thicker parisons and less for the thinner ones.

As obvious from the diagram, all outer dimensions of the die barring its length, are dictated by the configuration of the cross head and are beyond the purview of the mould designer. Whereas the core is centred automatically when screwed onto the mandrel, the die has to be centred after mounting with the help of three or four adjusting screws. Eccentricity yields a parison of uneven wall thickness, which would curl towards its thinner side, producing a banana effect.

The core and the die should preferably end flush with each other. The parison tends to roll inwards if the core stands far back from the die face. The effect is reverse, if the core juts out. The parison will curl outwards and stick to the die. Similar behaviour may be observed if the die and core differ too much in temperature.

The die-set with cylindrical parallel gap produces a pipe with uniform wall thickness. Any axial movement of the core with respect to the die brings very little variation in the wall thickness. However, it is a simple matter to alter the diameter and the thickness of the extrudate by working on the die or the core.

The design discussed above is quite effective with melts having pronounced swell characteristics. Its major disadvantage is that it offers very little scope for varying the wall thickness during extrusion

2. Tulip Die (Fig. 12.2)

Starting with an internal diameter equalling that of the cross head, the bore of the tulip die converges to a near-straight hole, maintaining an almost constant gap to the mandrel of the cross-head. After a certain straight length, it converges again to the desired parison size. The total angle of the die may vary between 50 to 30 degrees and that of the core within 30 to 20 degrees. The core may project a few millimetres beyond the face of the die.

The tulip die is particularly suitable for parisons of small diameters employed for internal calibration of the container neck.

3. The Trumpet die (Fig. 12.3)

As shown in the diagram, the bore of the die begins with that of the cross head and converges almost parallel with the mandrel

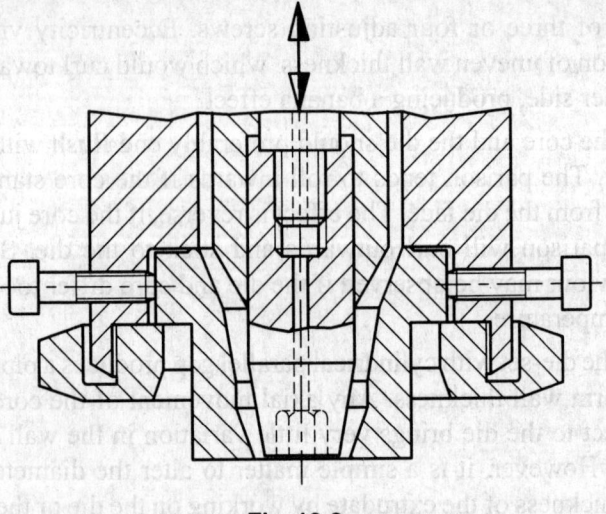

Fig. 12.2.

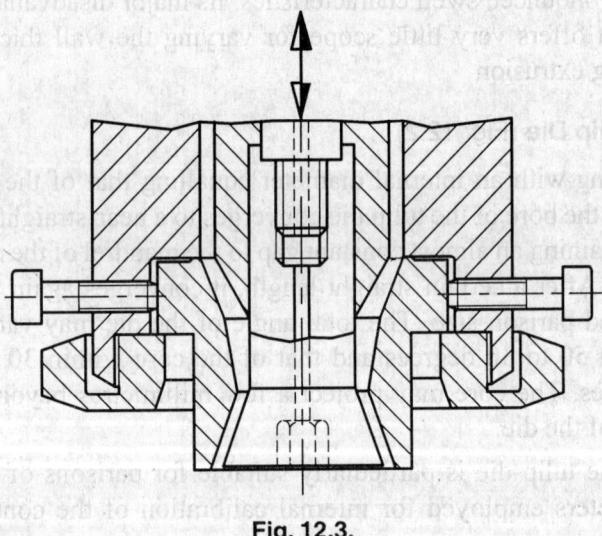

Fig. 12.3.

of the cross-head. After a certain near-straight length, the bore of the die diverges to the diameter of the parison. The core provides a constant gap for the thickness of the extrudate. The unchanging gap provides stress relief in the melt which, in turn, has a beneficial effect on die swell and the weight of the parison. The total included angle of the die should not exceed 40 degrees.

The trumpet die enables production of large, thin parisons, particularly suitable for blowing from below.

Parison Programming

Most blow moulded products have varying cross section which leads to varying stretching of a uniformly thick parison. Consequently, the wall thickness of the product is less at sections of higher stretching and vice versa. A product with varying wall thickness is prone to be weaker at thinner points. It also suffers from internal stresses due to uneven shrinkage. The remedy is to make a sectional adjustment in the wall thickness of the parison according to the degree of expansion.

All modern blow moulding machines are equipped with the so-called system of parison control, that is, of varying the thickness of a parison at predetermined junctures during extrusion. It consists of an arrangement to change the position of the core with respect to that of the die, usually with the help of a pneumatic cylinder, regulated electronically (Fig.12.4). With the dies and cores as described before, this leads to an increase or decrease of the annular gap between the two, which determines the thickness of the parison.

The method varies the parison thickness around the circumference at predetermined sections and is quite effective with round, symmetrical containers. But oval or unsymmetrical objects require thicker walls on one or more sides longitudinally. Several methods have been developed to this end.

a) Die Profiling

The orifice of the die is widened at required points to increase

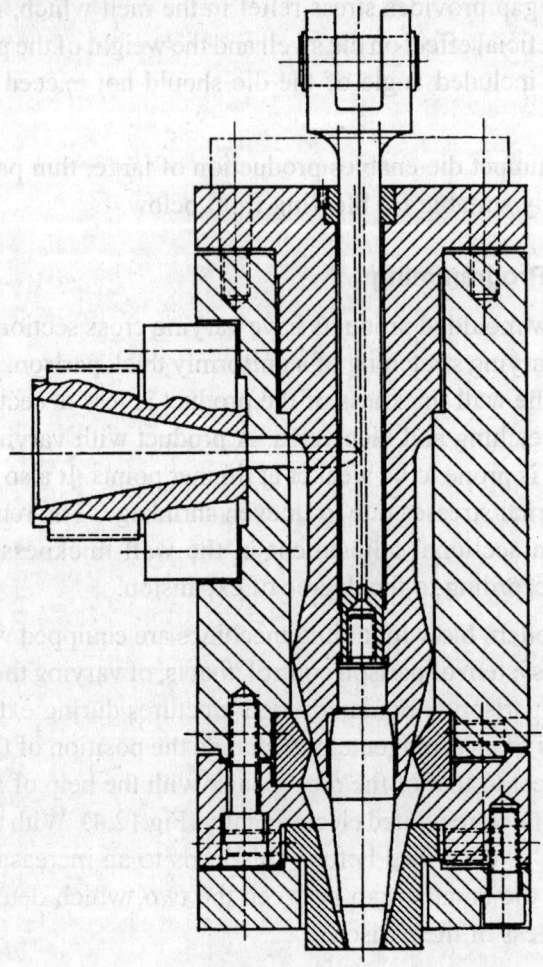

Fig. 12.4.

the wall thickness of the parison longitudinally (Fig. 12.5). It is an experimental approach, allowing thickness variation within narrow limits. Too thick a parison along one or more points may result in curling of the parison due to uneven extrusion. The alteration can also be applied to the core.

The die/ core altered in this way will serve only one particu-

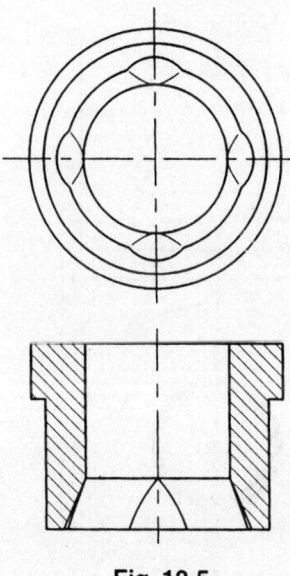

Fig. 12.5.

lar article. The change in wall thickness applies to whole length of the parison.

b) Die with Radial Movement

Industrial articles such as the air and fluid pipes in automobiles with 3D configurations may call for one-sided thicker walls in particular sections such as bends, curves and areas undergoing greater expansion etc. The method of radial wall thickness variation has proved more satisfactory in such cases. Here, the die is moved radially with respect to the core, creating an eccentric gap with thicker wall in the desired position. The die can be moved continuously to vary the wall thickness and its position in conformity with the requirement of the article (Fig. 12.6). It must, however, be realised that the radial movement beyond a limit can cause problems of parison curling due to faster extrusion on the thicker side. The radial movement is restricted to about 0.5 mm.

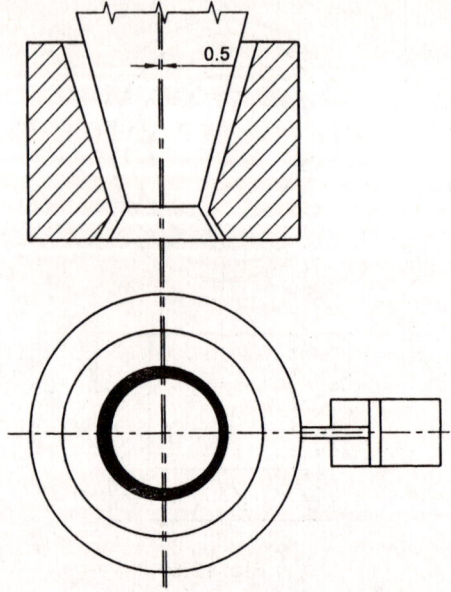

Fig. 12.6.

c) The Partial Wall Thickness Regulation

Profiling of the die to increase wall thickness of the parison at selected sections longitudinally as discussed before, serves a particular case. An improved variation, which can be adjusted and used universally, is the die set with a statically and flexibly deformable ring (SFDR). As shown in Fig. 12.7, the outer contour of the core is formed by a flexible ring, which can be deformed by setscrews to decrease the annular gap and thus decrease the thickness of the parison along selected sections. The ring is made of spring steel, rendered highly elastic through a special heat treatment.

The method is practicable for dies larger than 100 mm in diameter to accommodate the setscrews.

A further development in parison programming is the partial wall thickness regulation system, involving a dynamic, de-

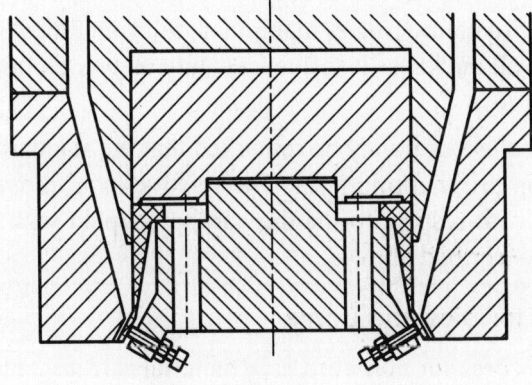

Fig. 12.7.

formable flexible ring (DFDR). Two or more hydraulic cylinders attached to the flexible ring, deform it as per the pre-set programme during extrusion of the parison, thereby changing the wall thickness sectionally. The cylinders can be actuated individually or collectively, at different times and to different extents. They pull as well as push so that the wall thickness can be decreased or increased locally. The cylinders are also in a position to push the ring eccentric to the static core (Fig.12.8). Whereas the setting of the static ring remains unchanged all

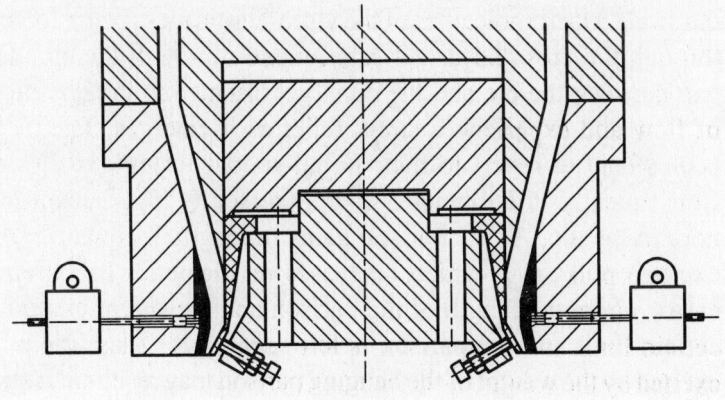

Fig. 12.8.

along, the dynamic ring is manipulated during the extrusion so that the wall thickness is influenced only on predetermined sections.

In conclusion, it may be remarked that the standard method of parison programming viz. The relative axial movement of the core with respect to the die can be coupled with all other systems described above.

Material for Dies and Cores

Dies and cores for non-corrosive materials are generally fabricated out of prehardened steels like P 20. This facilitates minor alterations and profiling. These can be hard chrome-plated for corrosive polymers such as PVC and POM etc. However, it is preferred to employ hardenable stainless steels for longer life.

Case hardening steels must be used for dies and cores when the plastics being processed contains abrasive additives.

All surfaces coming in contact with the melt must be smooth, free of machining marks and well polished.

The die swell

A very important but elusive feature, which influences the size of the parison besides the dimensions of the die and core, is the die swell. Macromolecules of the visco-elastic melt, being forced through the cross head and subsequently through the annular gap between the die and the core, get oriented in the direction of flow and experiences stress relief as it emerges. The molecules tend to revert to their initial state, causing expansion. Consequently, the diameter of the parison is bigger than the bore in the die. As the parison gains in length, its own weight exerts a pull on it, which decreases its diameter. In extreme cases, where the length of the parison and its weight exceed a certain limit and the parison is left sagging for long, the pull exerted by the weight of the hanging parison may result in "necking", decreasing the diameter and the wall thickness of the

parison close to end of the crosshead where it is most suscep-
tible to deformation because of higher temperature.

The extent of "die swell" is dependent upon a variety of
parameters viz. the material, its melt flow index, the speed of
extrusion, the length and weight of the parison and the
temperatures. Various studies have led to the conclusion that
unless the processing conditions are fully known and maintained
constant, it is very difficult to predict a reliable figure. The raw
material manufacturers too quote it as a broad range. For
instance, the swelling for a particular grade of High Density
Polyethylene may vary from 25 to 100 per cent. The values
observed under actual working conditions prove to be a more
reliable guide.

In conclusion, it may be pointed out that the swelling is comparatively less with diverging dies than with the converging
ones. Even if the annular gap may be decreasing with a diverging die set, the total area will be increasing towards the exit. In
other words, the melt has been experiencing relaxation even
before it leaves the die.

Calculating the parison diameter

It cannot be emphasised enough that any calculation of the
parison diameter can only be tentative. As explained above, the
die swell, the weight and length of the parison and some other
processing parameters, which cannot be predicted exactly before practical trials, exercise a far reaching influence on the
ultimate size of the parison.

For a round container, the base seam represents the starting
point of the calculation (Fig. 12.9). The folded parison should
be as broad as the base seam. Taking Lb as the seam length, the
outer diameter of the parison, Dp, works out as:

$$Dp = 2 \times Lb/\pi$$

The orifice diameter of the die can be derived by applying
the known or expected die swell factor. The wall thickness of

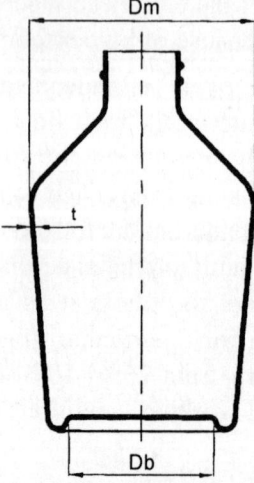

Fig. 12.9.

the parison with the calculated diameter Dp can be computed by considering the maximum diameter of the container Dm and the minimum wall thickness tm prescribed. The parison with the diameter Dp and the wall thickness tx is finally blown to the diameter Dm with a wall thickness tm.

$$Dp \times tx \times \pi = Dm \times tm \times \pi$$
$$tx = (Dm/Dp) \times tm$$

The diameter of the core is arrived at by subtracting twice the wall thickness of the parison, again making allowance for the die swell.

For containers with handles, it is assumed that the folded parison should cover at least half the breadth of the handle. In such cases, the distance of the middle of the handle from the centre line provides the basis for calculation of the parison diameter (Fig. 12.10).

It is a common practice to manufacture a few die sets closer

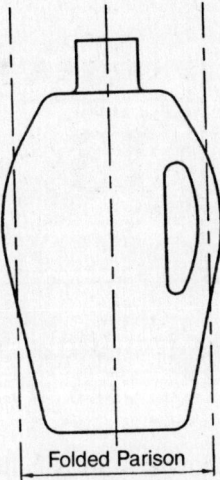

Folded Parison

Fig. 12.10.

to the calculated one to provide for the changes caused by unknown factors. Most workshops stock a complete range of dies and cores.

Materials for Blow Moulds

Each component of a blow mould is called upon to perform a specific function. This is why, different parts are fabricated out of different materials. The function dictates the choice of the most suitable material. Of course, other factors like the fabrication process, tool life, cost etc. play no less a role. However, a property common to all components coming in contact with the parison is a good thermal conductivity. Before selecting the right material, a mould designer should, therefore, familiarise himself with different materials, their properties and the methods of fabricating parts out of them. It is proposed to discuss in this chapter the suitability of various materials for mould components in the light of above guidelines.

1. Base Plate

Base plates are invariably flat and serve the purpose of mounting the mould on the machine platen. The compressive forces they are called upon to withstand are of low order. Mild steel has proved to be the right material for this component.

2. Guiding Members

Guide pins and guide bushes must be very accurate and should withstand wear and tear. Case hardening steel is the right material for these components. Hardened and ground, it gives them accuracy, a hard surface and a tough core.

3. Neck and Base Inserts

In spite of the fact that steel is a poor conductor of heat and that neck and the base constitute areas of heat concentration, the inserts for these sections of the mould must be made out of hardenable steel. It is here that welding of the open ends of the parison and cutting of the flash takes place. An idea of the wear and tear these are subjected to can be formed from the fact that the hardened base and neck inserts of a mould, capable of producing a few million mouldings, have to be replaced or repaired after every 100,000 shots. The slender cutting edges, if not renovated in time, yield a thick parting line, do not cut the flash cleanly and make deflashing a difficult proposition. The steel used should be of a through hardening type to permit corrections and regrinding. Hardness between 58-60 Rc is recommended.

4. Mould Cavity

Moulds for small and medium sized articles of regular geometrical shapes are generally machined out of unalloyed medium carbon steels. The poor heat conductivity of steel is compensated by its adequate tensile strength to some extent as the cooling can be placed fairly close to the cavity without impairing the strength and life span of the mould. Steel possesses a grain structure, which imparts a good surface finish to the blown product. The cavities are seldom required to be hardened unless the cutting edges run along the sidewalls.

Mould cavities for processing of PVC require a corrosion proof surface. These moulds may be machined out of medium carbon steel blocks and subsequently hard chrome-plated. It must, however, be borne in mind that the layer of chromium can peel off after sometime. It is, therefore, advisable to restrict the use of chrome-plating to the moulds, which are not required to produce large quantities. For mass production moulds, it will be found economical in the long run to manufacture cavities out of stainless steel.

An Aluminium alloy (DIN 3.4365), delivered in the form of pre-hardened blocks and plates, combines many advantages, especially for construction of blow moulds. It is light, adequately hard and tough but easily machineable and a very good conductor of heat. Its low specific gravity facilitates easy handling of even large moulds. Its excellent tensile strength endows the moulds with fairly long life and it has not been found necessary to incorporate steel inserts for cutting edges in the moulds for medium sized series. In other words, the complete blow mould can be machined out of two solid plates. The cooling effect is vastly superior to that in the moulds consisting of various materials with different rates of thermal conductivity. The alloy is capable of taking up polish adequate for industrial articles. The surface hardness can be further improved by hard chrome-plating, chemical nickel-plating etc.

It cannot be denied that machining moulds out of solid blocks is an expensive and a time consuming process, especially in case of large moulds. The quest for more economical mould manufacturing methods has led to the exploration of all possible materials and their suitability for blow moulds. The ensuing description should throw light on their merits, drawbacks and areas of application.

A. Casting

i) Steel Castings

Cast steel moulds cannot be regarded as finished tools as the cavity surface does not possess the finish and accuracy needed for the job. Further machining is therefore, unavoidable. A saving grace of the method is the fact that the castings can be provided with channels at the back, which can be used for circulation of cooling water. Sand castings are employed only in exceptional cases.

ii) Aluminium and Aluminium alloys

The low tensile strength and wear resistance of aluminium is amply compensated by its excellent heat conductivity, low specific gravity and good machinability.

A number of alloys has been specifically developed for casting of blow moulds. Improved casting techniques help further to overcome problem of blowholes often encountered in the process of casting.

Blow moulds of medium and large size are generally manufactured by casting of aluminium alloys. The low tensile strength of aluminium makes it necessary to use hardened steel inserts at all points of pinch off.

The low melting temperature of aluminium enables insert casting of the cooling network consisting of copper pipes bent according to the configuration of the article. Then process saves the cost and provides a very efficient and uniform cooling.

Aluminium is one of the cheapest mould making materials.

iii) Zinc alloys

Although costlier than the aluminium alloys, the zinc alloy castings are free from blowholes and other surface blemishes encountered in castings. The excellent reproduction of lines and texture of the casting model make subsequent machining superfluous. The cooling network, too, can be insert cast, as with aluminium castings.

Zinc alloys are very sensitive to pressure even at low temperatures. Great care is therefore, called for during handling and operation of the zinc moulds. The mould life can be considerably prolonged by the incorporation of steel inserts at critical points.

iv) Beryllium Copper Alloys.

Be-Cu alloys have proved to be ideal material for construction of blow moulds. Their distinguishing features are:

- Excellent heat conductivity
- Adequate tensile strength
- Hardenability
- Freedom from blowholes
- Good reproduction of intricate details
- Dimensional accuracy
- Corrosion proofness

Blow moulds cast out of Be-Cu alloys require very little or no surface finishing operation, depending upon the optical quality of moulding aimed at. Cooling pipes can also be insert-cast during the casting process as in the case of other light alloys. Even the pinch off can be formed out of the parent metal, as it possesses adequate hardness (~ 35 Rc).

The grain structure of Be-Cu alloys permits a high degree of polish. Chromium plating is, nevertheless, recommended for moulds meant to process PVC. The alloy enables good adhesion of the chromium layer, which may be as thin as 0.007 to 0.015 mm.

Sand casting is the cheapest, and in most cases, a sufficiently accurate method of forming blow moulds out of the Be-Cu alloys. Ceramic casting process, although elaborate, reproduces the finest surface details such as the wood grain, leather and textile appearance or any other structure. Pressure casting renders very high accuracy of dimensions coupled with exact reproduction of textural details, rarely needed for blow moulds.

The comparatively high cost of the Be-Cu alloys restricts their use to intricate moulds requiring faithful reproduction of textural details and very high accuracy of dimensions. A designer can bring down the cost by judicious combination of Be-Cu and steel components in the mould, employing the expensive metal for only those components where the advantages would set off the high cost.

v) Brass and Bronze

The use of brass and bronze is restricted to moulds of simple shapes requiring no pinch off along the sidewalls. Both alloys are good conductors of heat. Their low tensile strength narrows down their scope of application in spite of the fact that they are considerably cheaper than the Be-Cu alloys. Brass is easy to machine and is often used to manufacture inserts for logos, trade marks, scripts etc.

vi) Low Melting Temperature Alloys

Certain Tin-Bismuth alloys can be cast or sprayed at low temperatures. This property, coupled with their negligibly low shrinkage, has been made use of to manufacture "instant" moulds by spraying the molten metal on a sample or a model made of plastics, wood, plaster, metal, thermosetting resin, leather etc. in a matter of hours. A special formulation permits their use around 200 degrees C.

As these alloys are very expensive, usually a thin shell of 3-5 mm thickness is sprayed with a special gun on a model placed in a rectangular frame. Now the shell is either backed up with reinforced polyester, metal powder or sand impregnated epoxy resin, plaster or the alloy itself by pouring it in fluid state with a ladle. The process enables incorporation of cooling by placement of copper piping on the back of the sprayed shell before reinforcement is applied. A mould half is ready.

The main advantage of the process lies in its quickness, simplicity and its ability of faithful reproduction of the finest details. The metal lends itself to very smooth finish too. The softness of the alloy restricts the life of the tool to a few thousand shots only. As a rule, such quick moulds are used to produce a few prototypes for trials or a small pilot series. The expensive alloy can be melted and used again.

vii) Thermosetting Resins

Polyester and epoxy resins, impregnated with metal powders, can be used advantageously to manufacture experimental blow moulds. It must be conceded that these moulds are short-lived and can be employed to produce only a small number of mouldings for trials during the development phase of a new product. Their importance is realised fully in case of products for new applications where the design can be finalised after a number of experiments.

Moulds from thermosetting resins are manufactured by pouring the prepared resin over a positive model made of wood, plaster of Paris, metal etc., positioned in a rectangular frame. It is also possible to include a cooling network made of bent copper pipe before casting of the resin. A metal frame and hardened base and neck inserts help prolong the service life of the mould.

The main advantage lies in the simplicity of the process. No special equipment is needed to fabricate such experimental moulds in a short time at a low cost.

Table 13.1. Materials for blow moulds

Material	Composition												Specific gravity	Tensile strength (N/mm²)	Hardness	Heat conduct KJ/m.h.c.	Forming process	Application
	C	Si	Mn	Mg	P/S	Cr	Be	Co	Cu	Al	Zn	Fe						
Mild St. S 2062	0.2	-	-	-	0.055	-	-	-	0.2-0.35	-	-	Rest	7.8	412-530	-	-	Machining	Back plates
M.C. St. En 9	0.45-0.6	0.1-0.35	0.5-0.8	-	0.06	-	-	-	-	-	-	Rest	7.8	677	201-255	92-168	Machining	Cavity block
Grey iron S 210-FG40	2.5-3.75	1.0-1.25	0.4-1.0	-	S0.06-0.12	-	-	-	-	-	-	Rest	7.2	363-392	241-320	-	Casting	Cavity block
Alloy St. En 31	0.9-1.1	0.1-0.35	<1.1	-	0.05	1.0-1.5	-	-	-	-	-	Rest	7.8	588-882	60-65 HRC	92-168	Machining	Base & neck inserts
Stainl. St. AISI 405	0.08	1.0	1.0	-	S 0.03 P 0.04	14.5	-	-	-	-	-	Rest	7.7	441-588	-	92	Machining	Cavity block for PVC
Al. alloy M6-IS 617	-	10-12	0.50	0.1	-	-	-	0.1	-	Rest	-	0.6	2.6-2.7	167-216	-	823	Casting	Cavity block
Al. alloy AlZnMgCu 1.5	-	0.4	0.3	2.5	Ti 0.2 Ti+Zr 0.25	0.23	-	-	1.6	Rest	5.6	0.5	2.8	480-540 MPa	130-160 HBW	130-180 W/m K	Machining	Cavity block
Cu-Be alloy ISI C 17500	-	-	-	-	-	-	0.4-0.75	2.4-2.7	Rest	-	-	-	8.7-8.9	785-883	-	664	Casting	Cavity block
Zn alloy Zamak	-	Sn 0.1	Pb 0.1	-	0.2	Cd 0.1	-	-	2.8	0.8	Rest	<0.1	6.8	529	90-100 BHN	378	Casting	Cavity block

Mould Making Techniques

Majority of blow moulds for mass production, especially those for containers upto 30 litre capacity, are generally manufactured by conventional machining methods. A tool room, fabricating blow moulds, uses all conventional processes of metal removal such as turning, milling, drilling, boring, grinding as well as the specialised ones like copy milling, spark erosion and jig grinding. The machines, however, have become more sophisticated, efficient and accurate. CNC technology has brought speed, accuracy, economy and reliability to all processes of metal removal employed for mould construction.

It cannot be overlooked that the process of Electrical Discharge Machining (EDM) has revolutionised some basic concepts and removed all constraints on the design and shape of the product. It has also brought flexibility in the design of moulds.

Metal removal proves a time-consuming and also an expensive proposition in case of large moulds. The specialised methods aim at saving time and cost, not only for large moulds but also for those, big or small, which have to be fast and cheap because of a limited number of pieces they are destined to produce. It may, however, be underlined here that some components of these moulds, especially those meant to be hardened, are invariably fabricated by means of conventional methods of machining.

1) Casting

Sand casting of steel is a process suitable for manufacturing large blow moulds. Its main advantage lies in eliminating the tedious task of removing a huge amount of metal. The surface finish has to be improved in most cases by post operations. Another advantage is the possibility of casting the cooling chambers at the back of the mould block right at the time of casting.

Inserts for neck and bottom, guiding elements and venting arrangements have to be made by conventional machining processes.

2) Thermosetting resin casting

Moulds meant for a limited production can be readily cast out of special epoxy and polyester resins, impregnated with upto 80% aluminium powder, which lends them strength, heat conductivity and polishability. The positive model for casting can be made out of metal, wood, plaster, leather, silicone and plastics resins etc.

The model is positioned in a casting frame; metal inserts, such as guiding elements, base and neck inserts etc can be positioned and fixed at this stage. The cooling network made of bent copper pipes can be placed before the preheated and degassed mixture of resin and the hardener is poured over the model. The resin mixture cures or hardens to a solid mould block in the course of a day. Both mould halves are cast in this fashion separately, taken out of the casting frames after curing, matched and assembled and finally heat cured. They can be protected from damage by steel frames or side bars. The castings can be machined in conventional manner.

The resin reproduces all details of the model faithfully. It can be highly polished. The resin has negligible shrinkage

(0,02%) having practically no influence on dimensions of the blow moulded product.

The cast resins have adequate surface hardness but low tensile strength. Fitted with metal inserts for critical sections and metal frames, they are capable of producing about 1/10th. of the quantity that a steel mould may deliver.

3) Metal Spraying

A very fast method of producing a blow mould cavity is the spraying of low temperature melting alloys of bismuth and tin with the help of compressed air onto a pattern possessing the reverse shape of the intended cavity. The pattern can be made of metal, wood, epoxy resin or even thermoplastics. In other words, even a sample can be used as a pattern.

The pattern is treated with a release agent and the molten alloy is sprayed on it till a shell of about 4-5 mm. thickness is formed. Though the metal may have a temperature of 200-400°C, it cools down to 40-50 degrees when it hits the pattern and settles down on it in the form of a non-porous layer. The shell is backed up by epoxy resin impregnated with aluminium powder to form a cavity block. It is also possible to incorporate cooling by placing bent copper tubes in position before pouring the backing material.

The bismuth alloys are comparatively soft. Consequently, the blow moulds out of Bi-Sn alloys are seldom used for regular production. However, the speed of mould making enables quick production of prototypes or small pilot series. Here too, the mould life can be prolonged considerably by incorporation of steel inserts and protective mould frames.

The bismuth-tin alloys are quite expensive but they can be salvaged and reused for new moulds.

4) Rapid prototyping or selective laser sintering

This is the latest process of forming an article or a mould insert with the help of a laser beam, which is directed and moved by 3-D data on fine metal powder coated with thermoplastics. The thin layer of plastics melts under the heat of the laser and the metal powder sticks together in the shape generated by the beam. Layer by layer, the desired article called a green part is created out of the metal powder. It is now subjected to heat, which sinters it into a solid part.

The powder contains 60% tool steel and 40% copper.

The inserts, which have hardness equalling that of the mild steel, are built in a steel bolster to form a mould for small series.

A remarkable advantage of the process is the speed of manufacture without any model or pattern and the facility of incorporating integral cooling network during prototyping. The mixture of steel and copper has a better heat conductivity than the pure steel.

15

Mould Cost Calculation

In most tool rooms, the job of cost calculation, or more precisely, of cost estimation of a proposed mould, is assigned to the design department. An experienced designer can estimate the size of a mould and envisage operations and the time involved in its fabrication fairly accurately. A designer is required to have sound knowledge, not only of the design principles but also of materials and methods, of converting his design into an economical and efficient tool.

The most reliable method of cost estimation is to prepare a rough sketch of the mould and then estimate the time required to fabricate each component, listing each machining operation separately. The cost may now be computed by evaluating the hours on each machine separately or, as is more common, to apply an average hourly rate computed for the tool room to the total number of hours. Similarly, the cost of the tool design, hardening and testing etc. are added to arrive at a fairly exact cost of the tool.

Most tool rooms standardise the repetitive components such as the guiding elements, the neck and top inserts, the cutting rings, the bottom inserts, the venting plugs etc. so that their cost is known with a fair degree of accuracy. Likewise, the time required for various operations involved in mould making can be estimated within close limits on the basis of collected experience. Consequently, the operation of cost estimation is not as time consuming as it may appear at the first glance.

Data required for Cost Estimation of Moulds.

The manufacturing cost of a mould comprises:

- Design and development costs
- Raw material cost, including the price of purchased components
- Cost of machining and fabrication in own shop and outside.
- Cost of assembly and matching.
- Cost of trials and modifications.

On the basis of the manufacturing cost, the sales department computes the sales price by adding the following:

- Overheads.
- Taxes, risk cover, profit and interest
- Cost of packaging, despatch and insurance

The following chart has been devised to help calculate the cost of blow moulds. It will also be found useful in comparing the estimated and the actual costs after the tool has been completed.

Mould Cost Estimation Chart

Inquiry No. _____
Date _____

Article _____
No of cavities _____

Drg. No. _____
Sheet _____

S.No.	Part Description	No. off	Material			Bench Work	Rough Machining	Turning	Milling	CNC Milling	Drilling	Surface Grinding	Cyl. Grinding	EDM	Heat Treatment	Remarks
			Size	Wt.	Cost											
															Total	
															Hourly Rate	
															Total Cost	

Material + Machining: _____

Design + Development: _____

16

Design of Blow Moulded Articles

Design of the blow-moulded articles is governed by the following factors:

- Functional requirements
- Properties of the raw material
- Peculiarities of the forming process
- Aesthetic features
- Cost Considerations.

The blow moulding process is primarily employed to manufacture hollow articles out of thermoplastics. These may be containers for packaging of diverse goods, domestic as well as industrial or technical components. For the packaging articles, the choice of the raw material depends upon the contents, environment and mode of storage and the cost. The opening and hence the closure too has to suit the contents which may be in the form of fluid, paste, powder, crystals etc. The shape of the container, however, is the net result of all factors. It may even be regarded as the touchstone of a design as it has to take into consideration functional and aesthetic features, processing conditions, material properties, and, finally the economic aspects too. A successful compromise, blending all factors according to their importance, may be regarded as the best product design.

Blow moulded containers are finding application in widely diverse fields. It is, therefore, not possible to lay down rigid rules for their design. The basic principles, however, provide guidance in design and help to avoid pitfalls in all cases.

A hollow container consists of the following parts:

Neck

The neck of a blow-moulded container is generally round in order to permit the use of round closures, such as plugs, screw caps, crown corks and similar devices. Although it is possible to blow mould necks of other shapes too, they are seldom resorted to, as the process becomes complicated and expensive. The most common mode of closure is a screw cap. The neck of a blow-moulded container is, therefore, provided with external threads. The start and end of the threads can be moulded correctly only if they do not lie directly on the mould parting line (Fig.16.1). It must also be remembered that flash is mostly unavoidable around the neck as the parison is usually bigger than the neck diameter. Continuous threads, especially of fine pitch, make it extremely difficult to deburr the parting line, which leads to an imperfect fit between the neck and the screw cap. The threads on the neck should, therefore, be interrupted at the parting line. The quality of the fit is not impaired if the interruption of the threads forms a tangent to the thread core on the parting line (Fig. 16.2).

The start of the thread should be at least the height of one turn below the top edge of the neck. If, however, it is intended to achieve leak-proofness by means of a plug inserted inside the neck, the threads should begin lower than the end of the plug. The inside of the neck is formed by the blowing mandrel and only the unthreaded part of the neck remains smooth, enabling leak-proof closure. Good sealing can also be achieved if the top of the neck is stepped down so that the threaded portion

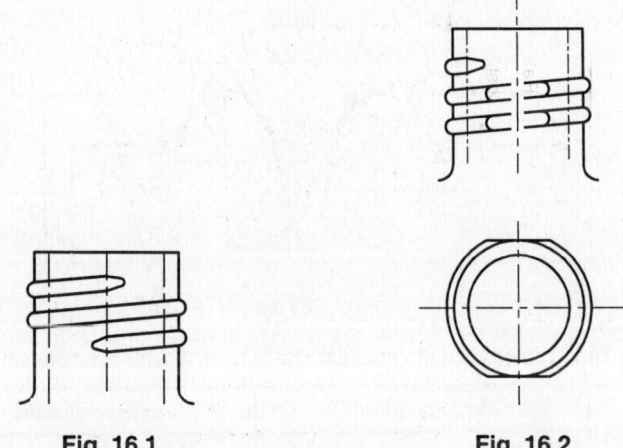

Fig. 16.1.   Fig. 16.2.

of it is independent of the straight part, which will remain smooth inside.

Threads

Round, trapezoidal and buttress threads have proved most suitable for blow-moulded articles. Round threads, though easiest to form, underlie a severe drawback. If the closure is screwed tightly, it may slip over the round flanks. Fine threads should also be avoided because of this reason. Unlike the threads cut in metals, the blown threads cannot be formed with sharp corners. All corners, especially the root of the threads, should be provided with radii.

Dimensions for round threads of different nominal diameters according to the German Standards (DIN 168) are given in Table 16.1

Closures

Blow-moulded containers can also be provided with closures other than screw caps. The devices used with glass bottles and metallic containers have proved successful here too, although

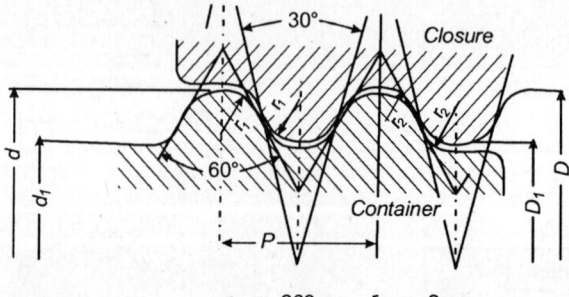

$$r_1 = .263\,p \; ; \quad r_2 = .2\,p$$

Table 16.1. Dimensions of container and closure threads for different dia.

Nominal dia.	Pitch 'd'	Container threads				Container threads			
		'd'		'd_1'		'D'		'D_1'	
		Max.	Min.	Max.	Min.	Max.	Min.	Max.	Min.
8	2	8	7.6	6.64	6.24	8.4	8.2	7.04	6.84
9	2	9	8.6	7.64	7.24	9.4	9.2	8.04	7.84
10	2	10	9.6	8.64	8.24	10.4	10.2	9.04	8.84
11	2	11	10.6	9.64	9.24	11.4	11.2	10.04	9.84
12	3	12	11.5	9.96	9.46	12.6	12.2	10.56	10.16
14	3	14	13.5	11.96	11.46	14.6	14.2	12.56	12.16
16	3	16	15.5	13.96	13.46	16.6	16.2	14.56	14.16
18	3	18	17.5	15.96	15.46	18.6	18.2	16.56	16.16
20	3	20	19.5	17.96	17.46	20.6	20.2	18.56	18.16
22	3	22	21.5	19.96	19.46	22.6	22.2	20.56	20.16
25	3	25	24.5	22.96	22.46	25.6	25.2	23.56	23.16
28	3	28	27.5	25.96	25.46	28.6	28.2	26.56	26.16
30	3	30	29.3	27.28	26.58	30.8	30.3	28.08	27.58
32	4	32	31.3	29.28	28.58	32.8	32.3	30.08	29.58
35	4	35	34.3	32.28	31.58	35.8	35.3	33.08	32.58
40	4	40	39.3	37.28	36.58	40.8	40.3	38.08	37.58
45	4	45	44.3	42.28	41.58	45.8	45.3	43.08	42.58
50	4	50	49.3	47.28	46.58	50.8	50.3	48.08	47.58
55	6	55	54	50.92	49.92	56.1	55.4	52.02	51.32
60	6	60	59	55.92	54.92	61.1	60.4	57.02	56.32
65	6	65	64	60.92	59.92	66.1	65.4	62.02	61.32
68	6	68	67	63.92	62.92	69.1	68.4	65.02	64.32
70	6	70	69	65.92	64.92	71.1	70.4	67.02	66.32
75	6	75	74	70.92	69.92	76.1	75.4	72.02	71.32
80	6	80	79	75.92	74.92	81.1	80.4	77.02	76.32
82	6	82	81	77.92	76.92	83.1	82.4	79.02	78.32
85	6	85	84	80.92	79.92	86.1	85.4	82.02	81.32
90	6	90	89	85.92	84.92	91.1	90.4	87.02	86.32
100	8	100	98.5	94.55	93.06	101.5	100.5	96.06	95.06
110	8	110	108.5	104.56	103.06	115.5	110.5	106.06	105.06
125	8	125	123.5	119.56	118.06	126.5	125.5	121.06	120.06
160	12	160	158	151.84	149.84	162.5	161	154.34	152.84
200	12	200	198	191.84	189.84	202.5	201	194.34	192.84

with suitable modification. For example, it is possible to use pilfer-proof tin screw caps on blow-moulded containers if the resilience of plastics is taken into account. The design of the neck does not differ from that used with glass bottles (Fig.16.3).

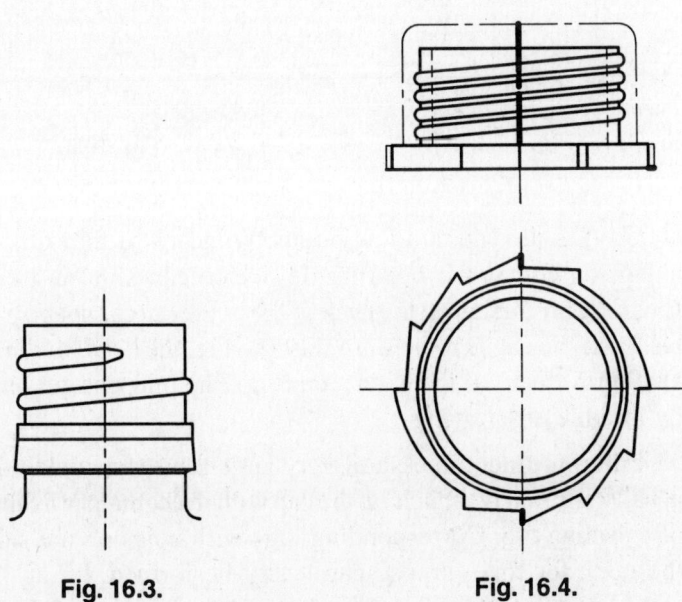

Fig. 16.3.   **Fig. 16.4.**

Apart from the conventional devices taken over from glass and metal containers, special configurations have been developed for plastics to achieve pilfer-proofness. One of them consists of a ratchet-like structure below the threads on the neck. Integrally connected to the plastics screw cap by means of a few thin ribs is a ring with corresponding teeth inside which can rotate over the ratchet in the direction of screwing but is prevented from rotating in the reverse direction due to engagement of the teeth. The unscrewing leads to severing of the ribs joining the ring with the screw cap (Fig.16.4). Another version used extensively with beverage bottles is a snap-on type screw cap having a ring as extension at the end (Fig. 16.5). The ring is

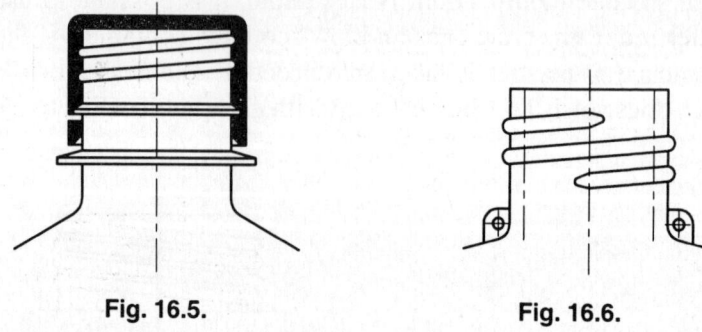

Fig. 16.5.   **Fig. 16.6.**

attached to the cap at a few points by means of thin ribs. The extension contains a bead inside, which can slip on a corresponding protrusion on the neck of the bottle but cannot be pulled back. As the cap is rotated for unscrewing, the bead hinders the axial movement of the extension ring. The thin ribs get broken by forcible unscrewing.

Pilferproofness in case of very large transport containers is usually ensured by attaching the cap with the container by means of a sealing tag. Corresponding lugs with holes on the side of the neck for this purpose can easily be formed during blow moulding (Fig. 16.6).

Shoulders

Shoulder is the section of a container, which joins its neck to the body. Horizontal shoulders are likely to distort on shrinking, which not only mars the appearance of the container but gives a tilt to the neck too. The remedy in such cases, viz., extending the cooling period, slows down the rate of production. It is, therefore, imperative that straight shoulders should be avoided. Shoulders slanting down from neck to the body retain a symmetrical appearance even after shrinkage. Although no specific rules have been laid down regarding a taper, a minimum angle of 10 degrees is recommended. It may be stressed upon that the junction of the neck with the shoulders and that of shoulders with the body must be generously rounded off. Especially the junction of the shoulders with the body forms a critical edge. Unless provided with a suitable radius, it may turn out to be very thin. The corner radius should not be less than $1/10^{th}$ of the container diameter if the blown article is cylindrical in shape. For products with an oval cross-section, this rule applies to the smaller diameter. Instead of a tapering shoulder, a convex radius also brings the desired effect. The radius improves the appearance too.

Body and Walls

Body is the part of the blow-moulded container which, in practice, has to withstand the maximum amount of stresses and strains. At the same time, it is here that the parison undergoes maximum stretching and if the shape is not nearly round, the product may exhibit here the maximum difference in wall thickness and consequently in strength. If other factors do not play a significant role, circular cross-section is the ideal one for blow-moulded articles. However, considerations like aesthetic appeal, storage space saving and stackability may make the choice of other shapes unavoidable in many cases.

As far as the blow moulding process is concerned, any shape can be produced, but drastic variations in length and breadth of cross-section lead to the vast differences in wall thickness. Rectangular and elliptical cross-sections may have a maximum ratio of 5:2 between breadth and depth or the major and minor diameter. In case of rectangular and square cross-sections, the corners should be well rounded to avoid over-stretching and excessive thinning of the parison. Flat faces tend to caving-in due to shrinkage. Slight ovality not only eliminates this defect but lends better appearance to the product too. As a thumb rule, a minimum ovality of 0.5-1 mm. for containers of 1 litre capacity and 1.5-2.5 mm for jerry cans of 5-litre capacity has yielded satisfactory results.

No hard and fast rules can be laid down for thickness of blow-moulded containers. The wall thickness depends upon various factors like shape, volume, rigidity of plastics material, mode of use, nature and weight of contents, method of storage and the cost. However, experience shows that generally a PVC/PET bottle of 3/4-litre capacity for packaging of liquids should have a minimum wall thickness wall of 0.25-0.3 mm and a bottle of equal capacity out of High Density Polyethylene should be 0.4-0.5 mm. thick at the thinnest point. The disposable containers form an exception.

Stiffness of containers can be increased by various means without increasing the weight. Ribs and beads in vertical or horizontal direction, gradually stepped up cross-sections and cross-ribs, depressed or protruding, are the usual design features resorted to. Various decorative designs like diamond shaped projections serve the dual purpose of reinforcement and beautifying. All such features should be kept shallow and well rounded to avoid over-stretching and thinning of the parison. The ideal cross-section for squeeze bottles is oval or rectangular. Resilience, a very important attribute for this function, is improved by convexity. Round shapes, too, can be employed to this end if the container is moulded out of flexible thermoplastics.

Handles and Grips

A necessity with containers of large sizes, grips and handles may also be sometimes required with bottles of comparatively smaller volumes. Basically, it is not difficult to blow-mould containers with integral handles provided the later lie within the range of the folded parison. A handle placed on the sidewall necessitates either a large parison covering, at least, the handle side of the article fully or employment of additional stretching devices. The result is a welding seam all along the vertical wall. A hollow handle is preferable to a solid one for the obvious ease of handling. The dimensions and location should be chosen with an eye on ease of moulding, convenience of grip and

Fig. 16.7.

harmony with the shape of the product (Fig. 16.7). The walls of the handle vertical to the parting line must be sufficiently tapered to facilitate ejection. The whole section of the handle should be generously rounded. It will, of course, be obvious that such handles have to be located symmetrically about the mould parting line.

Large drums generally require handling devices on two sides, diametrically opposite to each other. For occasional lifting, shell shaped depressions on walls suffice. (Fig.16.8). Unless they must be situated opposite to the parting line of the mould, the proper location for depression on the walls with undercuts is the mould parting line as the demoulding can be accomplished without additional devices.

Should it be necessary to provide separate plastics or metallic handles, solid lugs to fix the handles can be blow–moulded integrally. These lugs, situated on the mould-parting line can be twice as thick as the parison. It is obvious that the parison has to be sufficiently large to cover the lugs unless the part of the container having the lugs has been narrowed down. With

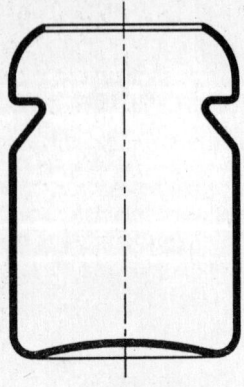

Fig. 16.8.

straight-walled containers containing lugs on sides, a welding seam along the parting line is unavoidable (Fig.16.9). It is also possible to insert-mould injection moulded lugs, especially when the thickness of the integrally moulded lugs (which can be twice as thick as the parison) proves inadequate. The process involves placement of these lugs as inserts in the blow mould. Adhesion is possible only when the lugs are moulded out of the same material as the container. The lugs should possess a thin but

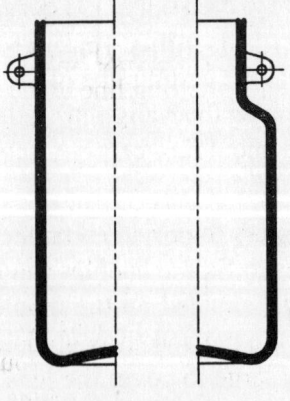

Fig. 16.9.

large base in the shape of a disc (Fig. 16.10). For safe placement in the mould, they should be located opposite the mould-parting line.

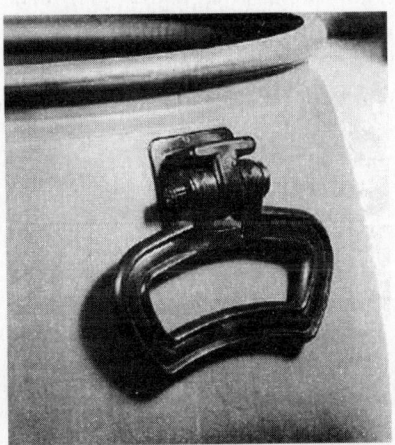

Fig. 16.10.

The base

The base of a container is required to provide a firm stay without rocking. As the end of the parison is pinched here, the base will always exhibit large differences in thickness and hence in shrinkage. The welding seam along the parting line may have maximum accumulation of material whereas the corners opposite to it may turn out to be the thinnest section of the container. The non-uniform cooling rate and hence the varying shrinkage will give rise to stresses, which if not taken care of in design, may lead to unevenness and even to failure of the product.

A flat base may give way under the stresses, which would effect its evenness adversely. Three or more spherical studs at the base may provide stability by eliminating rocking. However, blown studs, unless shallow, tend to be thin and may get punctured after sometime. Their use is, therefore limited to disposable containers.

A base, domed inwards, provides not only a firm stand to round containers but proves instrumental in releasing the stresses also. The concavity should start right after the corner radius. The camber height depends upon the diameter and volume of the container, wall thickness, stiffness and shrinkage of the material. A height of 1,5-2,5 mm. is quite common for bottles of upto 1 litre capacity (Fig. 16.11).

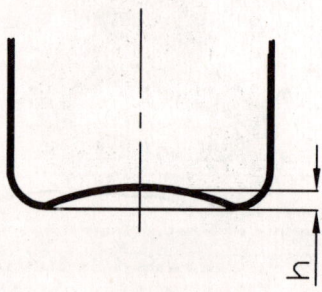

Fig. 16.11.

A roof shaped base serves the same purpose, but is generally employed for rectangular containers (Fig. 16.12). Here again, the camber height is decided by the factors mentioned already. For a 5 litre can, the following figures can be taken as guide values: flexible materials, 4-8 mm., semi-rigid materials, 3-6 mm. and rigid materials, 2-4 mm.

Camber heights exceeding these values pose difficulties during ejection. The base insert with the dome forms an undercut,

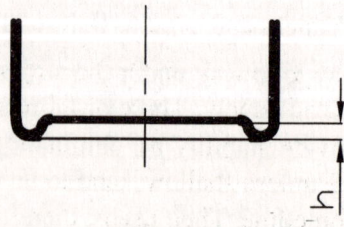

Fig. 16.12.

which may damage the articles as the mould opens. In case of large products where greater camber heights may be unavoidable, the mould design provides for the downward movement of base inserts for ejection of the product without deformation or damage (See also chapter 9, Undercuts).

A shell shaped depression in the base (Fig. 16.13) provides a good hold when large containers are tipped for pouring. The depression, located symmetrically about the parting line of the mould, does not form an undercut and the mould can open without deforming the product.

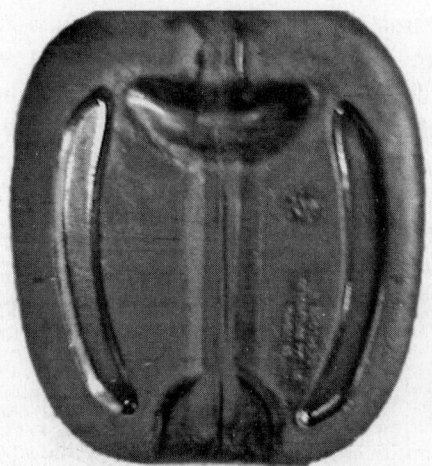

Fig. 16.13.

Inserts

The increasing application of blow-mould components for industrial purpose puts on them certain demands, which can be fulfilled only by inclusion of accurate and robust components like inserts. It is possible to incorporate metallic or plastics inserts in blow-moulded articles if care is taken that the inserts are well anchored and properly imbedded. Not only the position but also the shape of these inserts plays a decisive role. Metallic inserts do not undergo a bond with the blown container. It is only their shape and position, which can hold them in place securely. The location of inserts should enable plastics to envelope them on all sides. Their shape should ensure prevention of rotation. Fine ribs, sharp undercuts, teeth and knurling on inserts give rise to cracks in the plastics around them. Inserts with hexagonal sides have proved more satisfactory. Round inserts, with three or more properly rounded slots for anchoring, also yield good results.

Stresses are created when the plastics around metallic inserts is hindered in its normal shrinkage. In case of technical components subject to internal and external pressures and environmental strains, these stresses cannot be ignored. If the use of inserts is found unavoidable, the best way out is to injection-mould a sufficiently thick casing around them before incorporating them in blow-moulded body (Fig. 16.14). The plastics used in both processes should be identical. The injection-moulded casing not only serves as a buffer absorbing the stresses but also forms an integral bond with the container being blow-moulded.

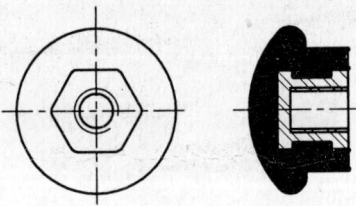

Fig. 16.14.

Certain features such as spouts, flanges, nipples, buttons etc., which may not come out well or may make the blow moulding process expensive, can also be insert–moulded. These injection moulded components form a sturdy bond with the blow moulded article, if they are pre-heated. This procedure combines the accuracy of injection moulding with ability of the blow moulding process to mould hollow articles with narrow openings. The product design should provide for at least one layer of the parison around inserts. Other rules regarding prevention against rotation and secure embedding must also be observed.

Post-fabrication

Insert moulding is associated with one drawback. The manufacturing process becomes necessarily semi-automatic and consequently slower. Automation is possible but economical only with very large series. The alternative lies in post-fabrication. In case of Polyolefins, additional components out of the same plastics can be successfully welded onto a blow-moulded article by application of heat and pressure, whereas cementing may serve the purpose with most other thermoplastics. The use of friction welding is confined to round components only.

Injection moulding can be employed to join parts made of dissimilar plastics by overmoulding a sleeve-like holder over the butted components. Anchoring devices like undercuts, flats, and slots add to the efficacy of the bond.

Film Hinge

An integrally moulded hinge can replace conventional hinges in certain blow-moulded products like toolboxes, packaging for sensitive and expensive instruments, insulated double-walled containers with integral lids etc., if the material used is polypropylene. The thickness of the hinge may vary between 0.1-0.3 mm. with a radius of 1.5 mm. The hinge should be preferably situated about 1.5 mm. below the face of the article so that it does not protrude when the container is closed. The measure guarantees a streamlined appearance of the product (Fig. 16.15).

Tolerances

Blow-moulded containers exhibit far greater variation in shrinkage and, consequently, in their dimensions, than the injection-moulded articles. The tolerances laid down by the German Standards Institute may help as a guide. The following tolerances apply to the blow-moulded containers made of polyethylene and rigid PVC:

Fig. 16.15.

Table 16.2. Tolerances for Blow-Moulded Containers

Volume	Upto 500 cm^3	Tol.	± 5%
Volume	500-2000 cm^3	Tol.	± 3%
Volume	2000-5000 cm^3	Tol.	± 4%
Volume	5000-20000 cm^3	Tol.	± 5%
Volume	20000-60000 cm^3	Tol.	± 6%
Volume	Above 60000 cm^3	Tol.	± 7%
Weight	Upto 5 gm	Tol.	± 10%
Weight	5-10 gm	Tol.	± 8%
Weight	10-15 gm	Tol.	± 7%
Weight	15-500 gm	Tol.	± 6%
Weight	500-1000 gm	Tol.	± 5%
Weight	1000–2500 gm	Tol.	± 8%
Weight	2500–5000 gm	Tol.	± 10%
Height	upto 30 mm	Tol.	± 3%
Height	30-100 mm	Tol.	± 2%
Height	100-250 mm	Tol.	± 1,5%
Height	Above 250 mm	Tol.	± 1%

Trouble Shooting

Moulding defects and their remedies

Trouble shooting may not seem to have direct link with the main subject of this book, viz. design of blow moulds, but viewing it from a different angle, it has to be conceded that as a mould wields considerable influence on the quality of a product, it will also be responsible for many defects. Trouble shooting, therefore may not be as foreign to design as it appears at the first glance. It is, in fact, as important for a blow mould designer to know the rules of design as the consequences of their non-observance. The malady has to be diagnosed from the symptoms of a bad product before a corrective action can be planned and undertaken.

The quality of a blow moulded article is the outcome of material properties, processing conditions, the mould design and its execution. Shortcomings in any one of these will be reflected in the quality of the final product. The question is rather complex. A defect may not emanate from only one source. In most cases, it will be found difficult to isolate a single cause of fault. However, before the responsible source is pin pointed and the course of action can be decided, it is essential to analyse the defect and identify its cause.

Although defects and difficulties in the production of a parison bear no relation with design of the mould, the knowledge of their causes puts the designer in a better position to understand the complete production process.

Table 17.1.

Defect	Cause	Remedy
One longitudinal streak opposite the melt inlet of cross head	Converging line caused by heart shaped curve or angular groove of the mandrel holder	Reduce cross section of the annular gap in cross head after melt has converged
Two or more thin longitudinal lines in uniform division on outer surface	i) Webs supporting the torpedo are very hot material at the joint stretches excessively	i) Reduce temperature of the cross head ii) Give webs a more favourable rheological design iii) Use a smaller head
One to several streaks, non–uniform distribution	i) The die has a rough surface ii) Old or degraded material sticking to the die	i) Clean and polish the die; provide about 0.5 mm radius at the mouth ii) Remove the old material and polish the die
Horizontal line on the outer surface of the moulding	i) Parison programming sets in too suddenly	i) Readjust the speed of parison control
Inadequate Blowing	i) Inadequate air pressure ii) Inadequate clamping pressure iii) Faulty/worn out pinch off inserts iv) Ejected article is very hot	i) Check air line for leakage; increase blowing pressure ii) Increase mould-clamping pressure iii) Repair/replace pinch off inserts. They should weld without cutting iv) Increase cooling time; improve cooling design

(Contd.)

Defect	Cause	Remedy
Blowing needle cannot penetrate the parison properly	i) Needle edge is worn out ii) Speed of penetration too low iii) Needle situated at an unfavourable point; parison is insufficiently supported at that place in the closed mould iv) Premature blowing	i) Replace/regrind needle ii) Shoot needle into the parison with higher speed iii) Shift point of penetration iv) Pre-blow the parison. Blowing through the needle should start after penetration
Article bursts during blowing	i) The parison has uneven wall thickness ii) The parison is too thin iii) Inflation is too slow iv) Pinch off is very sharp v) Cutting edges do not sit fully	i) Adjust die/ core ii) Increase gap between die and core. Increase screw speed iii) Increase air pressure iv) Round off the sharp edges v) Redesign cutting edges
Product deforms during /after ejection.	i) Premature ejection. The article is still very hot	i) Redesign cooling system in the mould ii) Cool the mould more intensely iii) Increase blowing period iv) Increase cooling period v) Decrease processing temperature vi) Introduce flush air-cooling
Product cracks or breaks upon ejection.	i) Notching effect ii) Melt overheated iii) Inclusion of foreign matter iv) Moisture in the material v) High internal stresses due to very low mould temperature	i) Redesign article; round off sharp edges, provide reinforcement ribs ii) Check and eliminate source of overheating iii) Clean screw, cylinder, cross head, die and core

Defect	Cause	Remedy
	vi) Stretching limit exceeded	iv) Pre-dry granules v) Raise mould temperature vi) Use a bigger parison
Article sticks to the mould.	i) Mould is too hot.	i) Redesign cooling ii) Increase cooling time
Incomplete reproduction of cavity details.	i) Insufficient blowing pressure ii) Duration of blowing is very short iii) Bad venting of the mould	i) Increase blowing pressure ii) Increase the effective blowing period iii) Improve venting
Uneven outer surface.	i) Poor venting design; air contained in the mould cannot escape ii) Excessive use of a mould release agent	i) Provide efficient venting ii) Roughen the cavity surface iii) Clean mould; do not use release agent
Rough Surface	i) Melt is too cold ii) Die has a rough finish	i) Increase the parison temperature ii) Decrease speed of extrusion iii) Increase die temperature iv) Polish the die
Dull finish adjacent to pinch off lines on the article	i) Damaged pinch off edges	i) Repair/renew the pinch off
Small round dots (tiny bubbles) on the article.	i) Raw material contains moisture	i) Pre-dry the granules
Large bubbles	i) Trapped air ii) Overheating of melt iii) pollution with oil in cylinder or cross head	i) Increase speed of screw ii) Install a screeni iii) eliminate source of leakage of oil
Mould parting line protruding (outwards) on the article	i) Mould is not closed completely ii) Cutting edges are worn out	i) Check mould halves and machine platen whether both are parallel

(Contd.)

Defect	Cause	Remedy
		ii) Increase mould clos-ing force
		iii) Renew the cutting edges
Mould parting line depressed (inwards) on the article.	i) Mould breathes ii) Premature inflation	i) Increase mould clos-ing force or lower blowing pressure ii) Start blowing after mould is completely closed
Article thinner at the end closer to the die.	i) Parison hangs too long ii) The temperature of parison is too high	i) Cut down the cycle time ii) Decrease tempera-ture iii) Use parison control
Excessive shrinkage	i) Blowing pressure is too low ii) Article is ejected when still very hot	i) Increase blowing pressure ii) Increase blowing time iii) Decrease parison temperature iv) Improve cooling de sign of the mould v) Increase cooling period
Poorly welded seam	i) Excessive clearance angle between the cutting edges ii) Breadth of cutting edge inadequate iii) Blowing starts too early or too late iv) Parison gets cold before welding	i) Decrease the clear-ance angle ii) Increase the breadth of cutting edges iii) Adjust starting time of inflation iv) Close mould without delay; increase tem-peratures; employ-core heating

Special Blow Moulding Techniques

The extrusion blow moulding process, as dealt with so far, may be termed as a universal method of manufacturing hollow articles out of thermoplastics. There is practically no limit to the shape or size of the object. Hollow articles of symmetrical or odd shapes, with or without openings, with handles, flanges, undercuts and inlaid inserts can be produced by this method. It is being successfully employed to produce miniature articles like vials of less than a cubic centimetre capacity to storage tanks of a volume over ten thousand litres. However, it cannot be denied that it has certain limitations, which may not be acceptable in certain cases.

The major shortcomings of the extrusion blow moulding process are:

- Uneven wall thickness and fluctuating weight.
- Hollow threads.
- Welding seam
- Flash

Although for general packaging requirements, these drawbacks may not be found disturbing, they may have adverse effect in more demanding cases. Special blow moulding processes, some of which will be described below, have been developed to circumvent these shortcomings.

1) Robot Aided Parison Placement (Fig. 18.1)

Industrial blow moulded articles such as exhaust gas pipes, fluid channels and ducts in automobiles and certain hollow connecting components in household machines like dish washers, washing machines, refrigerators etc. are usually curved and bent in three planes. Their production in the conventional blow moulding process necessitates very large parisons to cover the bends and entails huge waste. More often than not, the weight of the flash exceeds that of the product by far. Over and above that, a disproportionately large extrusion unit has to be employed to produce the big parison. Maintaining a uniform wall thickness too is quite problematic. The articles have a weld line all around and post operations like deflashing are unavoidable.

Several methods have been tried to overcome these difficulties. Among others, the method of guided placement of parison has proved to be quite effective.

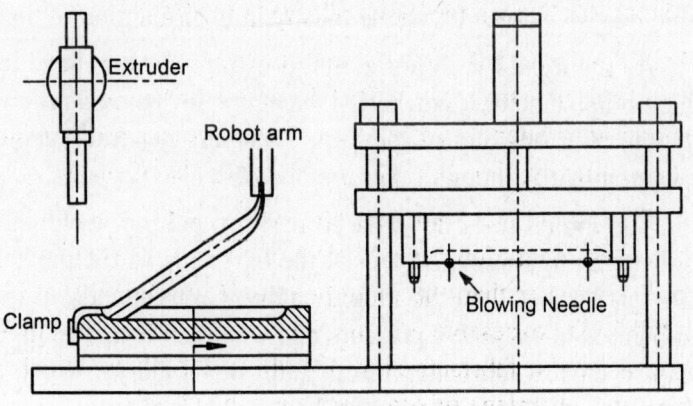

Fig. 18.1.

The Equipment

The blow-moulding machine is built around a horizontal extruder equipped with a cross-head. Seen from front, a vertical mould closing units is situated on one side of the output end of the extruder (Fig. 18.1). The upper platen is movable vertically. The lower platen is formed by a slide, which can move sideways towards the extruder on a bed. A robot arm is responsible for transport of the parison to the mould.

The Process

The mould closing unit opens and the lower half of the mould slides towards the cross-head on the horizontal bed. The robot arm grips the extruded parison (which is smaller in diameter than the smallest dimension of the mould cavity and is preblown with support air), severs it from the cross head and brings it to the mould half in a vertical hanging state. A special gripper provided at one end of the mould holds the lower end of the parison. The robot places the parison in the lower mould half, travelling in a programmed way following the contour of the cavity. The mould half returns, with the in-laid parison, to its place below the upper mould half.

Blowing commences in the closed mould with a hypodermic needle. A second needle may help in circulation of the air.

As the mould is in horizontal position, ejectors have to be incorporated in the lower half of the mould to demould the blown article after opening of the tool. These are generally actuated pneumatically, though other methods are also possible.

The mould does not have to perform cutting, welding and squeezing operations except at the two ends as the parison is always smaller than the mould cavity. Consequently, it is not subjected to excessive pressure and wear and tear. The moulds can be cast or fabricated out of light metal alloys, which also have the advantage of superior heat conductivity.

The machine can also have two mould closing stations, placed on either side of the extruder head.

Another variation of the method assigns the role of laying in the parison to the extruder itself, which is movable in two planes. The mould half, also in horizontal position, receives the parison from the extruder which moves in a pre-programmed fashion following the contours of the cavity. The end result is the same as described before.

It may, however, be pointed out that in both variations, only one side of the parison is in contact with the colder mould till it joins its upper half. It may disturb the heat balance in extreme cases.

2) Parison Suction Blow Moulding (Fig.18.2)

Parison suction blow moulding process serves the identical purpose viz. elimination of flash, unavoidable in conventional blow moulding of the 3-D articles. The objective is, however, achieved by sucking in the parison, which is smaller in diameter than the smallest cross section of the article to be blow moulded, in the closed mould.

The Equipment

The machine consists of a horizontal extruder equipped with an accumulator instead of a cross head at the output end. Built in the accumulator head is a blowing mandrel, which provides the support air to prevent the parison from collapsing. Two vertical platen, placed right under the accumulator and movable to and fro, form the mould closing unit. The unit is operated hydraulically. An important part of the equipment is the vacuum pump powered suction system, which reaches right under the closed mould and is joined to it after closing.

The Process

The mould consists of two halves containing the article cavity.

Phase 1

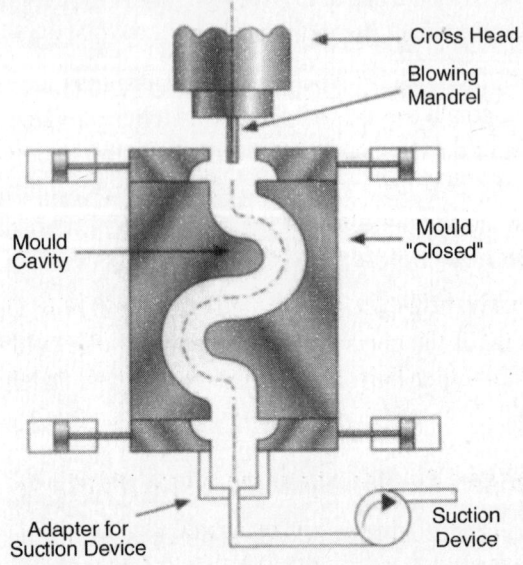

Phase 2

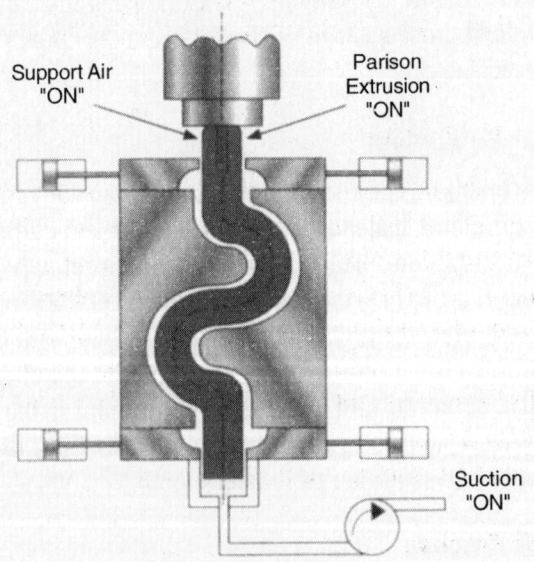

Fig. 18.2. Phase I & II.

Phase 3

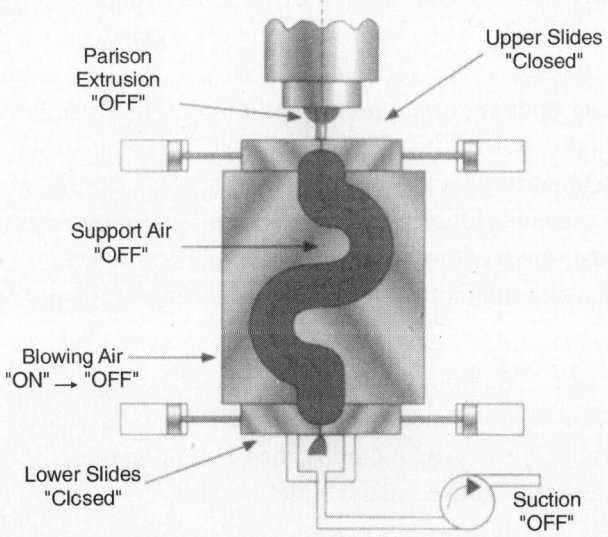

Parison Extrusion "OFF"

Upper Slides "Closed"

Support Air "OFF"

Blowing Air "ON" → "OFF"

Lower Slides "Clcsed"

Suction "OFF"

Phase 4

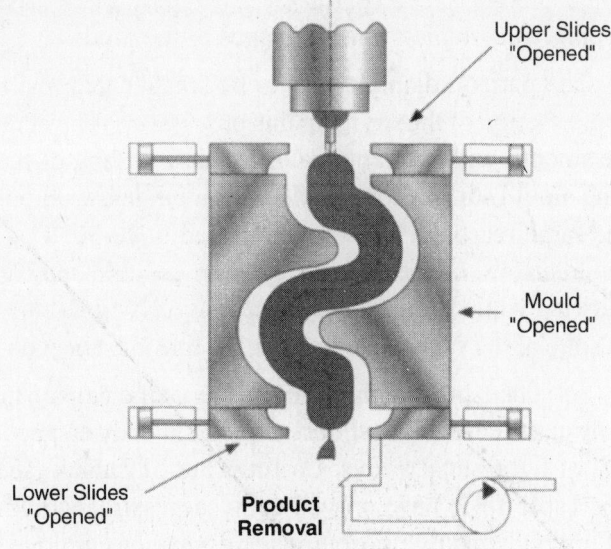

Upper Slides "Opened"

Mould "Opened"

Lower Slides "Opened"

Product Removal

Fig. 18.2. Phase III & IV.

It has one pair of slides on the upper end and another one at the lower end to close the two extremities of the open parison before blowing. The slides are actuated pneumatically or hydraulically (Fig. 18.2). The mould is closed. In closed position, the upper opening of the mould cavity lies directly under the accumulator head. The parison is expelled in one stroke under simultaneous operation of the parison programming and suction till the end of the cavity. Air in the mould prevents contact between the parison and the mould walls. Support air is also applied simultaneously to hinder collapsing of the parison.

The slides close and form domes at the two ends of the parison. The upper pair encloses the blowing mandrel jutting out of the extrusion head. The parison is inflated to the final form of the cavity, either by the blowing mandrel or by means of a needle housed in the slide.

As the mould opens after expiry of the cooling period, a pair of grippers holds the top flash and brings the article to a predetermined place, be it a conveyor or a bin, after the mould opens.

Fig. 18.2 depicts various stages of the process.

The parison diameter has to be smaller than the narrowest cross section of the cavity in this process too. The bends have to be smooth with generous radii for easy gliding of the parison. The mould surface too should be smooth and well polished for the same reason. The mould is heated to 80-90° C, so that the outer surface of the parison does not get cold and rigid before blowing while sliding through the mould. No gaskets are needed in spite of the suction but the parting line must be well matched.

As squeezing, welding and cutting of the parison takes place only at the ends in the sliders, the mould body can be fabricated out of light metal alloys. Castings out of Zamak (see Chapter 13, Table 13.1) have proved quite successful. Moulds cast out of impregnated thermosetting resins may be employed for pilot series.

The process has been developed and patented by KAUTEX Maschinenbau of Germany.

3) Injection Blow Moulding

Injection blow moulding process combines the advantages of injection moulding with those of the extrusion blow moulding and helps eliminate the negative factors listed before. Whereas in the extrusion blow moulding, the preform is always an open-ended pipe, the parison for the injection-blow moulding process is an injection moulded, closed bottom tube with a neck of the final shape and size. The wall thickness can be tailored to the shape of the final product. The preform has the accuracy and consistency, typical of injection moulding.

The injection blow moulding process can be carried out either continuously viz. from injection moulding of the preform till blow moulding of the article without interruption or in two stages, where the preforms are moulded separately, stored, reheated and then blow moulded independently.

The continuous process is illustrated in Fig. 18.3 in its various stages.

- The preform is injection moulded.
- The mould opens.
- The core is turned, transferring the preform to the open blow mould.
- The mould closes. Compressed air is introduced through the core and the preform is inflated to the shape of the cavity in the blow mould. The product is ready for ejection after cooling.

Though the injection moulding takes place with the melt in thermoplastic state, the blow moulding is carried out when the preform has cooled down to the thermoelastic level. The injection mould has to dissipate the heat in a controllable manner. The core fulfils four functions; it forms the inside of the pre-

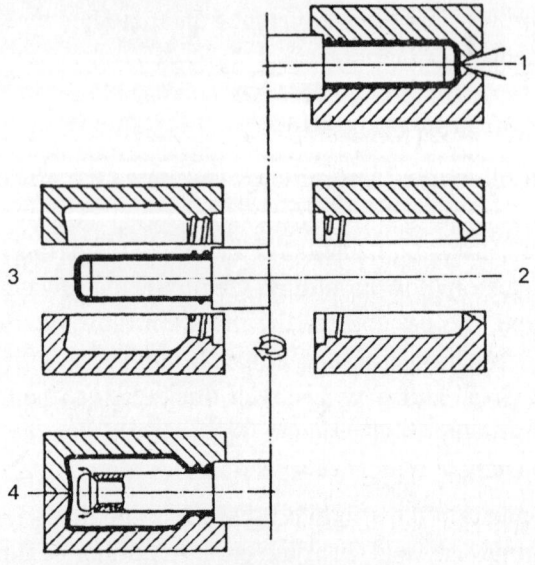

Fig. 18.3.

form, it acts as the carrier of the preform from injection to the blowing station, it cools the preform down to the thermoelastic temperature and it performs the function of a blowing mandrel.

Another variation of Injection blow moulding comprises stretching of the preform before blowing.

The injection blow moulding process has a number of advantages:

- Scrap-free production of hollow articles with seamless neck and base regions and high surface quality.
- High dimensional accuracy.
- Close tolerances.
- Uniform wall thickness.
- Solid threads
- Firm standing of the article due to absence of a base seam.
- Optimum transparency with amorphous polymers such as Polystyrene, Polyvinyl Chloride, Polycarbonate etc.

- Improved mechanical properties and better resistance to permeation of gases and water vapour due to biaxial stretching in the thermoelastic phase.

The injection blow moulding process has its limitations too. It is not suitable for unsymmetrical articles or those with additional openings, with hollow handles and lugs etc. This also holds good for containers with extreme cross sectional proportions such as rectangular bottles with breadth more than 2,5 times the depth or containers with narrow necks (limiting the size of the core) and very large bodies, necessitating blowing beyond safe limits.

The process may be carried out in line i.e. continuously or in two disconnected stages.

The continuous process calls for a special machine combining injection as well as blow moulding processes. The most common design has a vertical mould closing unit, a horizontal injection unit and the lower platen equipped with a turret in the centre (Fig. 18.4). The turret carries three identical cores which form the inside of the perform, cool it and blow it up. The process takes place in three steps:

1. Injection moulding

The injection moulding station has a horizontally placed injection unit, which injects the melt at the parting face of the two-part mould for the preform. The neck is formed to its final shape. The preform is cooled down to the thermoelastic stage before the mould closing unit opens.

2. Blowing

The turret lifts up and rotates by 120 degrees, bringing the core with the preform to the blowing station and an empty core to the moulding station. The mould-holding unit closes. The preform is inflated to its final shape. Simultaneously, a new perform is moulded at the first station.

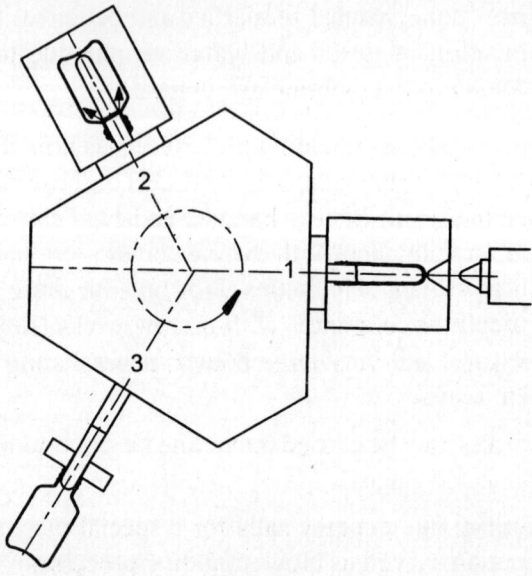

Fig. 18.4.

3. Ejection

The mould-closing unit opens and the turret executes a 120-degree rotation. The blown article is transferred to the ejection station, the freshly moulded perform to the blowing station and the empty core to the injection station. A fresh cycle starts after the unit closes.

The continuous process, also called the one stage process, is carried out in first heat. The most important factor here is the exact reproduction of the heat history from injection till blowing. It is vital for the quality of the product that all parameters be maintained in closest limits.

The process can be modified to include mechanical stretching of the preform before blowing. The biaxial stretching improves tensile strength, impact strength, clarity and the barrier properties.

The one stage injection blow moulding can also be carried out on a standard injection moulding machine. Here, the injection mould is equipped with a blowing station and a core transferring mechanism. Quite a few mould designs have been successfully developed. Fig. 18.5 shows one such design (See box).

The one-stage injection blow moulding process, though expensive in equipment, offers many advantages. These are:

- No intermediate storage of preforms
- No reheating before blowing
- Less space requirement
- Only one-time handling

The in-line injection blow moulding is quite successful in case of amorphous thermoplastics as they cool down to the thermoelastic phase readily. (The biaxial stretching arranges the otherwise haphazardly placed macromolecules and this endows the product with improved properties as mentioned before. The case is different with the semi-crystalline or crystalline polymers).

The two-stage injection blow moulding performs the two main operations, viz. injection moulding of the preform and its inflation, independent of each other. The preforms are moulded on a standard injection moulding machine and generally stored away. They are reheated to the thermoelastic stage before blowing.

The process may appear more expensive at the first glance but it has a definite edge over the continuous one. It does not call for a special machine and the cycle is not dependent upon the cooling time for the parison. The preforms are produced by multi cavity injection moulds, allowed to cool down and then reheated under precise control and automatically transferred to blow moulds where they are stretched and blown by the mandrels fitting in the opening. The two processes need not run simultaneously. Preforms for bottles of various sizes and shapes

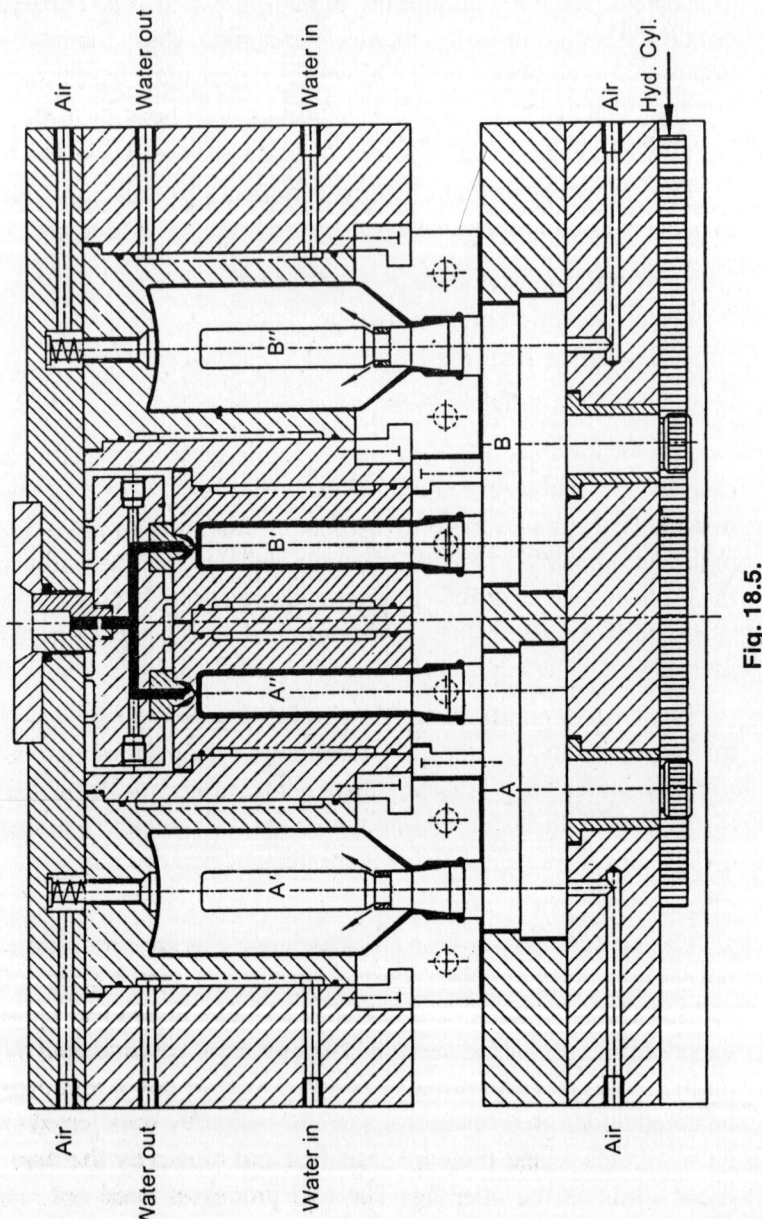

Fig. 18.5.

The two cavity composite injection blow mould for a 500 cc. bottle is designed to run on a 100-ton injection moulding machine equipped with a core pulling device to move the rack. The cavities for the injection moulded preforms are placed in the middle and those of the blow moulds on either side of these, all arranged in a row. Two revolving disc A and B, housed in the moving half of the mould, carry two preform cores each. The discs can be turned by 180 degrees by means of a hydraulically operated rack, engaging the pinions mounted on the spindles of discs. The rotation brings the cores with moulded preforms to the blow moulds and the cores from the latter to the injection moulds. The neck of the preforms is formed by a common slide, attached to the moving half of the mould, whereas the slides forming the neck of the bottles constitute a part of the fixed side.

Operation. The preforms are injection moulded and cooled down to the visco-elastic stage through the cooling arrangement provided in the cavities and cores. The mould opens. The preforms adhere to the cores. The rack rotates the discs and brings cores with moulded preforms in line with the blow moulds. Simultaneously, the empty cores are positioned opposite the injection moulds. The mould closes. New preforms are moulded in the middle and the preforms in the blow moulds are inflated by compressed air introduced through the cores. On completion of the cycle, the air is let out and the mould opens, leaving the blown bottles in the cavities. These are ejected by air ejectors. The slides are rotated again and a new cycle begins.

have been standardised and it is not uncommon that some moulders produce only the preforms and supply them to manufacturers of final containers.

The Preform

A distinct advantage of the injection blow moulding process is the possibility of tailoring the preform matching the final product in wall thickness distribution. The usual method is to work on the contour in the cavity and to keep the core straight but with a small taper so that unblown preforms can be stripped off easily during the setting up stage (Fig.18.6). Extreme wall thickness differences, however, may lead to moulding problems like weld line formation, air entrapment, uneven cooling and non-uniform shrinkage. Generally, the preforms for round, square and rectangular bottles have the shape of a test tube with threaded neck, a closed, spherical bottom and a fairly uniformly thick body. The wall thickness may lie between 2 to 5 mm. The blown article may display slight variation in wall thickness, which is generally of no consequence.

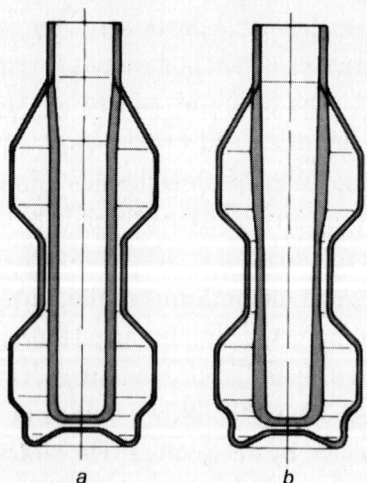

a   *b*

Fig. 18.6.

The diameter of the neck of the container to be blown exercises a decisive influence on the size. A core longer than 10-12 times the neck diameter is prone to deflection under the moulding pressure. The neck diameter also limits the container size, as the safe blowing ratio is 1:4. The length of the preform, if not to be mechanically pre-stretched, should be about 75-80% of the blown container.

The Moulds

i) The injection mould

In the continuous process, the injection mould invariably consists of two cavity halves. The neck portion should be designed as a separate insert in each mould half, as it needs intensive cooling. It may be pointed out that the neck section does not undergo any inflation after moulding. The rest of the mould calls for a separate well laid out cooling network, which may be provided with the cooling medium at a different temperature.

The mould body can be fabricated out of pre-hardened steels such as P20 when non-rigid plastics like low density polyethylene is being processed. Special mould steels, which can be hardened with minimum distortion—DIN 1.2767 is one such material—have proved better for more rigid thermoplastics. The moulds for corrosive polymers like PVC must be fabricated out of hardenable stainless steels. The neck inserts must be hardened in all cases. The stainless steel DIN 1.2083 (see also chapter 13, Table 13.1) is being employed for moulds of non-corrosive plastics too as it resists rusting in the cooling channels and is hardenable and polishable.

In order to prevent sticking of the moulding, the preform moulds are well polished and in most cases, coated too.

Medium carbon steels prove adequate for fabrication of the supporting structures like insert holding frame and clamping plates etc. Pre-hardened steels are more expensive but lend rigidity to the mould.

ii) The blow mould

The blow mould for the in-line injection blow moulding process underlies practically the same design rules, which are applicable to the extrusion blow moulds. However, these moulds are not required to have pinch off and the related features. Generally, the venting provided on the parting face proves adequate. A point worth emphasising is the perfect matching of the neck threads in injection and the blow mould. The thread contours in the blow mould may be 0.1-0.2 mm larger than those in the injection mould for the preform.

iii) The core (Fig.18.7)

The preform core for the in-line process may be considered in two parts as far as its design parameters are concerned. The dimensions of its section forming the inside of the preform are governed by the design of the parison and the shrinkage allowance for the polymer. It must have a minimum taper of 0.5 degrees for stripping off the unblown preforms during the setting-up stage. It has to be made up of two parts to permit inflow of the blowing air. The long cores are designed so that the blowing takes place near the top, just below the radius. This lends greater rigidity to the core. For containers with L/D ratio below 4:1, the parting of the core for air exit may be close to the neck of the container. The core needs intensive cooling to bring down the preform temperature from the moulding point to the thermoelastic one. Depending upon the thermoplastics, it may range from 65° to 135°C. The cooling cartridges have proved suitable for cooling of the long, round cores.

As the core has to withstand very high pressure, it must be made of a tough, hardenable material having good polishability.

The back portion of the core, which is held in the turret, has to conform to the configuration foreseen by the machine designer. It is here that the cooling water and the compressed blowing air are introduced in the core. Also the preform stretching

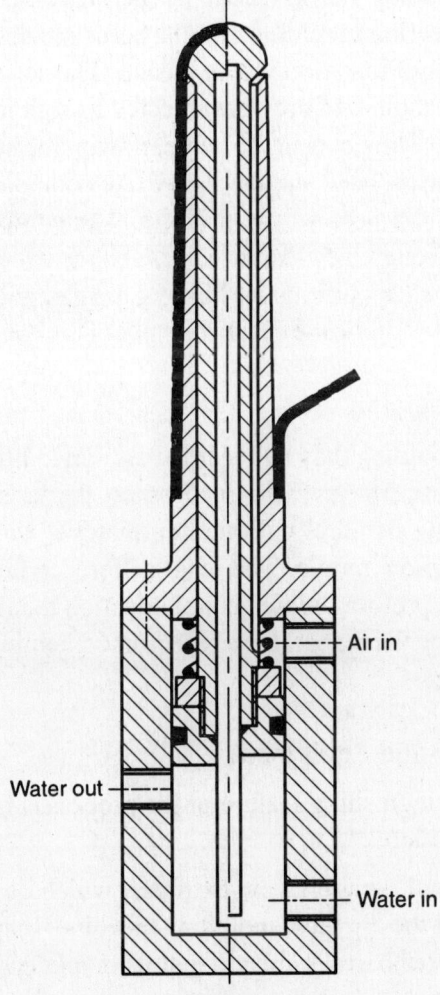

Fig. 18.7.

mechanism is housed in the turret. The design and dimensions of this section of the core may differ from machine to machine.

The cores for the preforms for two-stage injection blow moulding differ substantially from the foregoing ones in design and construction. The first stage of the process involves conventional injection moulding and the cores are designed as per the guidelines of injection mould design. The cores for stretching and blow moulding are not subjected to high forces or high temperatures. They consist of two parts viz. the stretching rod with the spherical head and the sleeve like component around it fitting inside the neck of the preform. The compressed air is introduced through the clearance between the two components. Here too, the back portions for holding the pair and moving the stretching rod underlie the design dictates of the equipment used.

Dip Blow Moulding

Dip Blow Moulding may be regarded as akin to Injection Blow Moulding in principle as in this process too, the parison is formed over a blowing mandrel with molten material and then transferred to the blow mould. The main difference lies in the formation of the preform, which is not injection moulded but created by dipping the mandrel in a reservoir containing plasticised thermoplastics.

The Process and Equipment

The most distinguishing feature of the process is its method of parison formation.

A horizontal, single screw extruder, which constitutes the central unit of the dip blow moulding machine, supplies melt to a vertical hollow barrel joined to it at its output end. The lower end of the barrel is closed by a piston capable of moving up and down whereas the upper end is equipped with an interchangeable die ring. A mandrel or "core", held by a pair of jaws on its upper end, dips into the barrel (Fig. 18.8 a). The jaws contain the contour of the neck of the container to be moulded. At b, the mandrel has dipped fully in the melt. As shown in C, the piston moves up and fills of the neck contour in the jaws. The mandrel

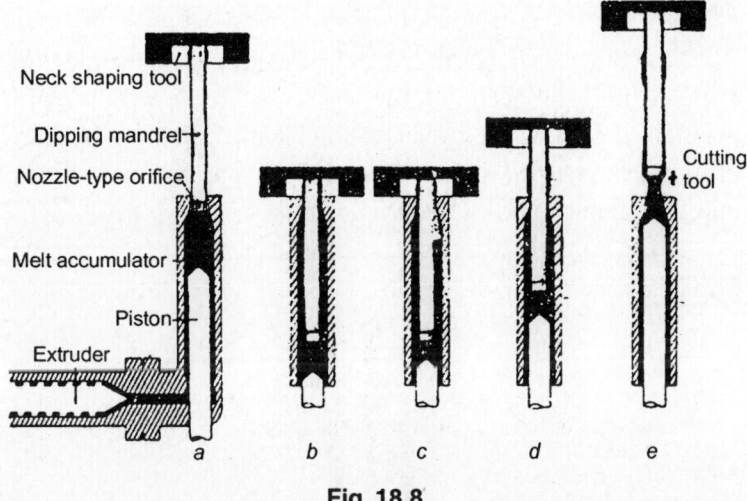

Fig. 18.8.

starts retracting, followed by the piston. The amount and thickness of the melt layer emerging out clinging to the mandrel depends upon the die gap as well as on the relative speed of the mandrel and the piston (d). It can be varied continuously. "e" depicts the final stage of parison formation, where the mandrel has emerged completely out of the barrel bearing a parison of varying wall thickness as pre-programmed. A knife cuts it free from the barrel.

Fig. 18.9 depicts a two-station dip blow moulding machine, with the mould closing units arranged on either side of the extruder. The vertically placed mould platen close horizontally. The moulds in the illustration are equipped with two identical cavities. The extruder supplies melt to two barrels.

The dip moulding process can be divided into four distinct phases, starting from dipping of the mandrel into the accumulator barrel till ejection of the cooled blow moulded product. The following description pertains to the two station machine with a double impression mould.

Phase 1

The blowing mandrels, embraced by the neck shaping jaws, are lowered into the accumulator barrels, which are filled with plasticised plastics. The jaws rest on the die ring after the mandrels are submerged to their entire length. The pistons move up and fill the neck part of the article housed in the jaws.

Phase 1

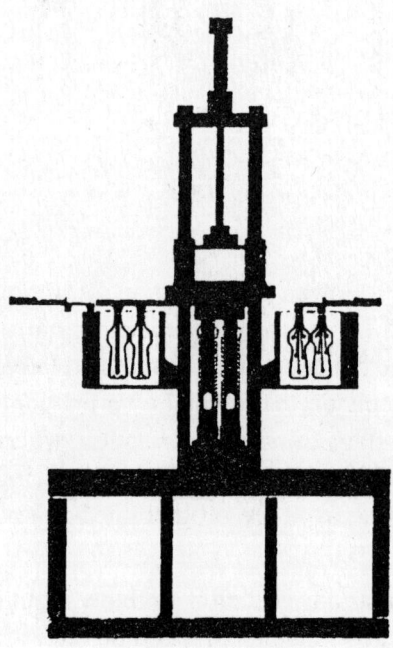

Fig. 18.9. Phase I.

Phase 2

After formation of the neck, the mandrels are pulled up gradually with the pistons following them with variable speed, forcing the melt through the die rings onto the mandrels. The thickness of the melt layer around the mandrel is proportional to the speed of the pistons, which can be varied continuously without interrupting the process. The annular gap between the die ring and the mandrel exercises no influence on the wall thickness of the preform. When the piston forces up more material than the receding mandrel is capable to pull out through the die ring, the thickness of the melt around the mandrel would exceed the annular gap. Conversely, the melt layer will be less than the clearance if the piston displaces less material than the mandrels are capable of pulling up.

Phase 2

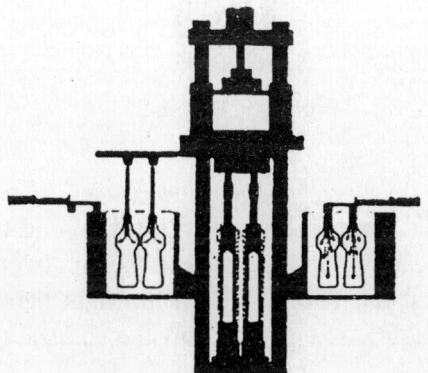

Fig. 18.9. Phase II.

Phase 3

The mandrel-bearing slide is lifted to its uppermost position, retracting thereby the melt coated mandrels out of the accumulator barrels and the other set on the left out of the blow moulds. A post-blowing fixture, that keeps up the air pressure in the blown bottles and ventilates them before ejection, replaces the latter.

The neck forming jaws open and the preform-bearing mandrels can now be transferred to the blow mould on the right hand where the mouldings from the previous cycle have cooled down and the compressed air has been let out.

Phase 3

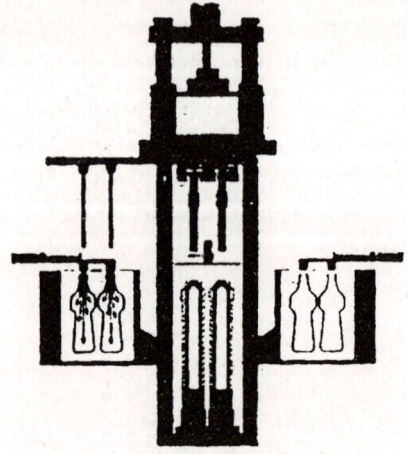

Fig. 18.9. Phase III.

Phase 4

The slide makes a stroke towards right, bringing the preform-bearing mandrels over the mould and the empty ones over the accumulator barrels. The mould on the right hand opens and the finished articles are ejected. Now the slide is lowered. The

Phase 4

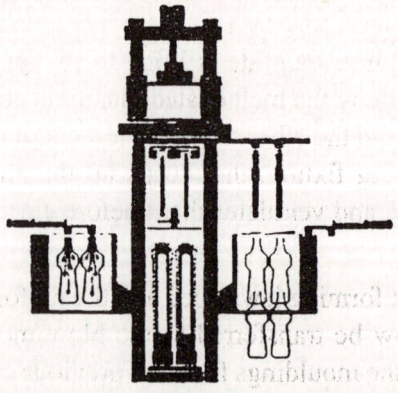

Fig. 18.9. Phase IV.

mould closes over the mandrels with preforms and the blowing operation commences. Simultaneously, the other set of mandrels dips into the accumulator barrels.

In the next cycle, the slide will move towards left and the same operation will be repeated.

Although the process appears similar to that of the injection blow moulding, it distinguishes itself by the lack of an expensive injection mould. As the mandrels are not subjected to a high pressure during formation of the preform, there is no danger of their bending. Consequently, the size of the neck of the article docs not impose any limit on its length.

Whereas any change in wall thickness calls for time consuming alterations in the injection mould or the core in case of the injection blow moulding process, the same can be effected in this process by a simple adjustment of the programmer.

As with Injection blow moulding, here too it is not possible to mould hollow bodies with integral handles.

The Moulds

As obvious, no injection moulds are required in this process for manufacturing the preform. Slides and mandrels are the only mould components, which shape the preform. The blowing mandrels are not subjected to high pressures and can be made out of any case hardening alloy steel yielding good surface finish. The mandrels must be provided with effective cooling.

The blow moulds have the function of giving the final outer shape and to cool the inflated article. As in the foregoing process, no cutting and sealing features are needed. The moulds can be made out of unhardened steel or preferably out of aluminium alloys offering the advantage of more efficient cooling. The mould halves can be fabricated out of single material blocks with inserts for the neck portion for better matching, more intensive cooling and interchangeability.

Designing a Blow Mould

1. Study the product design for its suitability for blow moulding. The criteria:
   Blowing ratio
   Corner radii
   Abrupt changes in cross section
   Tolerances
2. Check special features like:
   Wide mouth
   Additional openings
   Bosses on side
   Undercuts
   Inserts
   Grips and handles
   Holes and slots
   Texture
3. Choose the parting line.
4. Decide the position of the cavity in the mould. The criteria:
   Minimum diameter of the parison
   Minimum possible welding and flash
   Uniform stretching
   Trouble-free ejection

5. Decide the method of blowing.
6. Decide location of blowing.
7. Decide the mode of cooling.
8. Add shrinkage allowance on the article dimensions.
9. Add draft angle.
10. Locate air traps.
11. Decide position of bottom and top inserts.
12. Estimate the parison dimensions.
13. Calculate the locking force required and check the machine capacity.
14. Calculate the dimensions of flash pockets.
15. Draw the article in sectional elevation, as it would be blow moulded.
16. Draw the lines where top and bottom inserts will be placed.
17. Draw plan of the article in section below the elevation.
18. Draw cooling holes/chambers in plan and project in elevation.
19. Draw cooling of top and bottom inserts and finalise their size.
20. Decide the position of guiding elements in Elevation.
21. Finalise the length and breadth of the mould in Elevation.
22. Finalise the size of the back plates.
23. Draw cutting edges and flash pockets.
24. Draw air vent plugs and venting slots.
25. Locate screws and dowels.
26. Locate holes for eyebolts for lifting.

BIBLIOGRAPHY

Bibliography

1. R.Holzmann, Development of blow moulding technique and the production of hollow articles. Kunststoffe 10/61.

2. D.Hofman. The production of hollow pieces from Makrolon. Plastverarbeiter 10-12/63.

3. R.Holzmann. Verschluesse fuer Kunststoff-Flaschen. Verpackungsschau, 6/66.

4. H.Pelka. Production of large sized articles from Makrolon on blow moulding machine with melt reservoir. Plastverarbeiter 10/67.

5. P.Bauer, BASF. Spannungskorrision beim Polyaethylen. Kunststoffe 11/67.

6. W.Wilborn. Formgestaltung und Falllfestigkeit von Flaschen aus PVC hart. Verpackungs-Rundschau 11/1968.

7. Walter Hellerich. Kunststoffe. Franksche Verlagshandlung, Stuttgart, 1968.

8. BASF. Kunststoffe in Konstruktion, Werkstoffblatt 3222.1S 1-12, August 69.

9. R.Holzmann. Das Blasformen thermoplastischer Kunststoffe. Maschinenmarkt 1969.

10. Dipl. Ing. A. Schneider BASF. Blasformen von Kunststoffen. Stand der Technik, Probleme und Tendenzen. Kunststoffe 10/ 69.

11. H.Frank, BASF. Material- und Fertigungsgerechte Gestaltung von Kunststoffen. Verpackungsrundschau 5/70.

12. Dipl. Ing. W. Preddoel, Dipl. Ing. G. Wuebken. Einfluss verschiedener Herstellbedingungen auf die Eigenschaften geblasener Hohlkoerper. Plastverarbeiter 8/70.

13. A. Thomas & W.Kramer. How to eliminate faults when blow moulding hollow articles?

14. W. Mink. Grundzuege der Hohlkoerperblastechnik. Zechner + Huetig Verlag.

15. R. Holzmann. Gestalten von Blasformteilen. VDI Verlag.

16. Dr. Ing. Stoekert, Dipl. Ing. Dominghaus,.O.Plajer & E. Ulrich. Formenbau fuer Kunststoffverarbeitung, Carl Hanser Verlag.

17. Hoechst Plastics. Blow moulding of thermoplastics.

18. BASF. Lupolen, Verarbeitungstechnik, Band II.

19. O. Plajer. Werkzeuge fuer das Blasformen, Z+H Verlag.

20. Kunststoff Taschenbuch. Carl Hanser Verlag.

21. Bayer AG. Verarbeitungshinweise, Novudur.

22. R.Holzmann. Fortschritte beim Verpacken mit Kunststoffen-Extrusionsblasen und Spritzblasen. Industie-Anzeiger 2/71.

23. H.Frank, BASF. Polyolefine fuer das Blasen von Behaeltern. Industrie-Anzeiger 3/71.

24. E.G.Fisher. Blow Moulding of Plastics. The Plastics Institute, 1971.

25. Dipl. Ing. W. Predoehl, Dipl. Ing. W.Pflueger. Der Kraftbedarf der Bodenschweissnaht beim Extrusionsblasen. Plastverarbeiter 10/71.

26. W.R. Bursian. Vom PVC-Pulver zum Verpackungs-hohlkoerper—Produktionsprozess und Problemloesungen im Blasbetrieb. Plastverarbeiter 10-11/71.

27. Gulf Oil Company, U.S. High Density Polyethylene.

28. Gulf Oil Company, U.S. Blow Moulding Trouble Shooting Chart.

29. Gulf Oil Company, U.S. Poly-Eth. HI-D Linear Polyethylene.

30. I.C.I. England. The Blow Moulding of Welvic PVC. Technical Note W 110.

31. I.C.I. England. Alkathene; the Blow moulding of Polyethylene, Technical Note A 106.

32. Mining & Chemical Ltd. England. The Swift method of Tooling for Plastics. Products.

33. Lonza, Switzerland. PVC-Compounds fuer Verpackungszwecke.

34. Shell Chemicals, USA. Blow moulding Polypropylene.

35. BASF, Germany, Kunststoffverarbeitung im Gespraech-3, Blasformen 1973.

36. R. Schuhbach. Optimierung des Quetschkantenbereiches und des Schliessvorganges fuer Blaswerkzeuge. Plastverarbeiter 10/73.

37. Brydson & Peacock. Principles of Plastics Extrusion. Applied Science Publishers London, 1973.

38. Hans Dominghaus. Die Kunststoffe und ihre Eigenschaften. VDI-Verlag, 1976.

39. Technologien des Blasformens. VDI_Verlag. 1977.

40. DIN 168. Rundgewinde.

41. DIN 6094. Kronenkorkmundstuecke.

42. DIN 6131. Kanister aus Kunststoff fuer fluessige Fuellgueter.

43. German Patent 971333. R. Hagen & N. Hagen. Verfahren und Vorrichtung zur Herstellung von Flaschen und aehnlichen mit einer Einfuelloeffnung versehnen Hohlkoerpern aus thermoplastischen Kunststoffen.

44. BASF. Technische Information, Lupolen 4261 AX. Verarbeitungshinweise fuer das Hohlkoerperblasen.

45. BASF. Technische Information, Ultramid B6, Verarbeitungshinweise fuer das Blasformen.

46. Erich Gruber, Dr. Dietrich. Polymerchemie. Steinhoff Verlag, 1980.

47. Blasformen von Polypropylen. VDI-Verlag. 1980.

48. German Patent DE 3413201 C1, B.Braun Melsungen AG, Erfinder Batra Ramesh; Spritzblasvorrichtung. 1985

49. Frank/ Biederbick. Kunststoff-Kompendium. Vogel Verlag. 1988.

50. Donald & Dominik V. Rosato. Blow Moulding Handbook. Hanser Publishers Munich, 1989.

51. E.Mack & H. Schaefers. Arbeits- und Pruefungsbuch, Kunststoffverarbeitung. Vogel Verlag, 1989.

52. Steffi Baetz. Blasformen—Einflussgroessen bei der Voroermlingsbildung. Dr. Reinold Hagen Stiftung Bonn, 1995.

53. Normn C. Lee. Blow Molding Design Guide. Hanser Verlag, Munich. 1998.

54. A.R.Shekhar, Du Pont India. Trouble shooting in blow moulding. Popular Plastics and Packaging, 1998.

55. Schwarz/ Ebeling/ Furth. Kunststoffverarbeitung. Vogel Verlag. 1999.

56. Blasformen '99. VDI Verlag, 1999.

57. Otto Schwarz. Kunststoffkunde. Vogel Verlag, 2000.

58. Norman C. Lee. Understanding Blow Moulding. Hanser Verlag, Munich 2000.

59. H.G.Elias. An Introduction to Plastics. Wiley-VCH, 2003.